Conditioning

Wanda Wyrwicka

Conditioning

Situation Versus Intermittent Stimulus

Transaction Publishers
New Brunswick (U.S.A.) and London (U.K.)

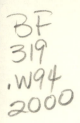
Library of Congress Catalog Number: 99-16172
ISBN: 1-56000-432-0
Printed in the United States of America

Library of Congress Cataloging-in-Publication Data

Wyrwicka, Wanda.
    Conditioning : situation versus intermittent stimulus / Wanda Wyrwicka.
      p. cm.
    Includes bibliographical references and index.
    ISBN 1-56000-432-0 (alk. paper)
    1. Conditioned response. 2. Psychology. Comparative. I. Title.
BF319.W94   1999
152.3'224—dc21   99-16172
                                                           CIP

Contents

Acknowledgments

I wish to extend deep appreciation to my colleagues, Drs. Michael Chase, Carmine D. Clemente, Ronald M. Harper, Rebecca K. Harper, and Barry M. Sterman, for reading parts of the manuscript and for their most helpful comments. Special thanks are due to Dr. Irving Louis Horowitz, chairman of the board and editorial director of Transaction Publishers for his excellent suggestions and especially for his generous help with selecting the title of the book. I also wish to thank the publishers of the University of Chicago Press for permission to reproduce some of the illustrations used in this book.

I also owe very best thanks to my daughter Joanna-Veronika Warwick for her valuable editorial assistance.

W.W

Introduction

Traditionally, in laboratory research, the process of conditioning, both classical and instrumental type, is initiated with a single intermittent stimulus (such as a tone, flash of light, etc.). Because of its role in evoking conditioned behavior, the use of an intermittent stimulus has become an indispensable part of laboratory research on conditioned behavior.

A question arises whether the same scheme of conditioning may be applied to behaviors occurring in real life. Let us analyze some cases from human behavior. One example is illustrated by the sound of an approaching plane. If this event happens in wartime, this sound can evoke the defensive reaction of escaping to a shelter. But in peacetime, the sound of a plane usually evokes only a slight orienting reaction.

Another case, less dramatic than the first mentioned is an event observed everyday in school. A bell sounding before class results in students entering the classroom; the same sound at the end of the class results in students leaving the room. Therefore, the same sound of a bell evokes two different behaviors, depending on the circumstances in which it occurs.

In each of these examples, the single stimulus (the sound of an approaching plane or ringing of a bell) does not determine the reaction; instead, the decisive factor here is the *situational background* against which the single stimulus appears.

A question can be asked—what happens when the situational background is suddenly changed? In a case related to me, a young married couple invited the wife's mother, who was still living alone in the "old" country, to live with them in the United States. The mother came, but could not adjust to the new life-style and became openly very unhappy. On her request the daughter took her back to her home in the old country. There are similar cases among new immigrants to this country. One immigrant told me, "there is no life here" and constantly returned to talk about his earlier times "there."

Similar observations were made on domestic pets. Some animals cannot easily adjust to new conditions of life. I know of a case where a dog was given to an older couple who lived about 100 kilometers from the animal's previous home. One week later, looking very exhausted, this dog managed to find its way back to its former masters. The animal had probably not eaten anything during the several days of its trip. So deeply touched were its masters, they decided to keep the dog.

In the laboratory, reactions to situational stimuli are often disregarded, and their occasional occurrence is sometimes treated as a phenomenon requiring a new theory.

This book discusses various cases of situational influences on behavior and attempts to provide a proper explanation.

Chapter 1 briefly describes some of I. P. Pavlov's historic achievements in creating a theory of conditioned reflex, a basis for understanding all behavioral activities of the organism from a physiological point of view.

Chapter 2 presents results of studies by Konorski and Miller on motor (instrumental) conditioned reflexes showing various kinds of these reflexes and some differences from the classical Pavlovian conditioned reflexes.

Chapter 3 deals in full with early data on the phenomenon of so-called "switching" obtained in studies of Konorski and Miller and in the laboratory of Asratyan.

Chapter 4 describes the effects of changes in the experimental situation on the value of conditioned reflexes.

Chapter 5 consists of two parts. The first part presents results of studies on the role of the experimental situation in instrumental conditioning; the second part describes the studies on the single (intermittent) conditioned stimulus tested in various situations.

Chapter 6 discusses examples of complex instrumental reflexes such as detour behavior and activities related to food intake and sleep.

Chapter 7 narrates cases of conditioning of activity of internal organs.

Chapter 8 shows the therapeutic role of EEG feedback in epilepsy.

Chapter 9 contains a theoretical discussion of the data described in previous chapters. It emphasizes the crucial role of situational background in determining the final results of conditioning procedures.

Rounding out the book, a general summary and conclusions once again emphasize the effect of the situation on conditioning processes.

1

Classical Conditioned Reflexes

Ivan Pavlov's historic discovery of the principle of the conditioned reflex was an enormous break in the scientific approach to the understanding of brain functions.

One of Pavlov's first observations in this area was made during a study on the secretion of saliva in dogs. Namely, Pavlov found that the animals salivated not only when they were eating food offered by a caretaker, but also when they saw the caretaker approaching or heard his footsteps. This simple observation indicated that the inborn reaction of salivation to food (or other objects) in the mouth may also appear to originally neutral stimuli which accompanied or immediately preceded the food intake.

As a physiologist, Pavlov chose a physiological interpretation of this phenomenon (Pavlov 1926, 1960). He accepted the concept of the *reflex* introduced by Descartes (1764) that "an external or internal stimulus falls on some one or another nervous receptor and gives rise to a nervous impulse; this nervous impulse is transmitted along nerve fibers to the central nervous system. There, due to the existence of nervous connections, it gives rise to a fresh impulse which passes along outgoing nerve fibers to the active organ where it excites a specific activity of the cellular structures" (Pavlov 1960, p.7). Pavlov considered the secretion of saliva to the presence of food in the mouth (and specifically to the tactile and taste stimuli produced by the morsel) an inborn reflex and called it an *unconditioned reflex* (UCR) while the salivation observed to the stimuli previously neutral (such as e.g., sounds of the caretaker's footsteps) he called a *conditioned reflex* (CR). The stimulus that produced unconditioned reflex was

5

called an *unconditioned stimulus* (USC). The previously neutral stimulus that produced conditioned reflex was called *conditioned stimulus* (CS) (e.g., a definite sound or a visual object). This terminology has been widely accepted by researchers.

Pavlov emphasized that the establishment of a conditioned reflex is dependent on the application of the unconditioned stimulus (which is also called the *reinforcing stimulus* or simply *reinforcement*) during or following the action of the conditioned stimulus; it is necessary that the conditioned stimulus starts to operate *before* the application of the reinforcing stimulus. If the conditioned stimulus is not followed by the unconditioned stimulus, the conditioned reflex cannot be established.

According to Pavlov, "conditioned reflexes are phenomena of common and widespread occurrence: their establishment is an integral function in everyday life. We recognize them in ourselves and in other people or animals under such general names as 'education', 'habits' or 'training'; all of these are really nothing more than the results of an establishment of new nervous connections during the post-natal existence of the organism. They are, in actual fact, links connecting the extraneous stimuli with their definite response reactions" (Pavlov 1960, p. 26).

In Pavlov's laboratory a great variety of conditioned stimuli were used. These included various sounds (called by Pavlov "stimuli of the acoustic analyzer"), visual objects ("stimuli of the visual analyzer"), touch ("stimuli of the tactile analyzer"), various smells ("stimuli of the olfactory analyzer"), changes in environmental temperature ("stimuli of the thermal analyzer"), and even the sensory feedback produced by performance of a passive movements ("stimuli of the motor analyzer"). Practically each change in the environment could become a conditioned stimulus on condition that it could be noticeable by the corresponding receptor. For instance, a thermal stimulus of 45°C applied to the dog's skin could be made into a conditioned stimulus, whereas a stimulus of 38–39°C, similar to the usual temperature of the dog's skin, was ineffective in spite of the fact that food reinforcement was offered after each application of this stimulus.

Manipulations with a variety of conditioned and unconditioned stimuli resulted in an accumulation of a great amount of experimental data that served as constant enrichment of the Pavlovian theory. Let us reminiscence with some topics of study in Pavlov's laboratory (Pavlov 1960).

One of these topics was the process of inhibition of the previously acquired behavior. Inhibition could be either external or internal. External inhibition could be caused by an appearance of an extraneous stimulus, not involved in the conditioned reflex arc. For instance, such extraneous stimulus could be a sudden noise, not used previously, an unknown person who appeared suddenly during the presentation of the conditioned stimulus, etc.

Internal inhibition is strictly related to the already formed conditioned reflex and occurs *inside* the arc of this reflex. A simple example of internal inhibition is a case when the reinforcing stimulus (e.g., food) has been withheld. When this procedure occurs repeatedly in the following trials, the conditioned reflex (e.g., salivation) gradually diminishes and finally stops appearing to the conditioned stimulus. This phenomenon was called by Pavlov the *extinction of the conditioned reflex* (Pavlov 1960, p. 49). It could be obtained even during a single session when successive presentations of the stimulus remained nonreinfored. However, the conditioned reflex to this stimulus recovered as soon as the reinforcement was given again.

Internal inhibition observed in Pavlov's laboratory included a form called the *conditioned inhibition.* This type of inhibition was obtained when the conditioned stimulus was preceded by a new stimulus and, in this case, the reinforcement was not given. After several repetitions of this manipulation, the conditioned stimulus, preceded by the new stimulus, never produced the conditioned salivation whereas the conditioned stimulus applied alone did (Pavlov 1960, Ch.V.).

Another case of internal inhibition studied in Pavlov's laboratory was the "*inhibition of delay.*" This kind of inhibition consisted of extending the time interval between the beginning of the action of the conditioned stimulus and the reinforcement. The prolongation of the period of the delay could be achieved gradually when (after establishment of a conditioned reflex with a usual reinforcement given 1–3 seconds after the beginning of the action of CS) the reinforcement was delayed by an additional, e.g., 5 seconds, in each consecutive session, until the delay reached several minutes. As a result of such procedure, the conditioned salivation did not start immediately after the onset of the conditioned stimulus, but only 30–60 seconds or longer after the beginning of action of the CS. The same result could be achieved when after the establishment of the conditioned reflex the delay in reinforcement was prolonged to the desired length rapidly, without any gradual steps. However, in this case, some complications in the dog's behavior could appear during the training period (Pavlov, p.89).

Finally, a form of internal inhibition was obtained when two similar conditioned stimuli were used and one of them was always reinforced by food whereas the other was not. As a result of this manipulation, the conditioned salivation was present only to the CS which was reinforced each time, whereas it was absent to the other stimulus. This form of internal inhibition has been called *differential inhibition* or *differentiation* (Pavlov 1960, pp.117–125 and following).

It should be mentioned, however, that the above described differential inhibition does not mean a complete disappearance of the reflex. In some circumstances such as, for instance, the entrance of a new person into the experimental chamber, the previously inhibited reflex may be disinhibited and the conditioned salivation may reappear (Pavlov 1960, p. 115).

Generalization of Stimuli

An important phenomenon observed mainly during the initial phase of establishing the conditioned reflex is the tendency to react not only to one particular stimulus but to all stimuli similar to it. For instance, when a conditioned reflex has been established to a tone of specific frequency by reinforcing it with food, many other tones will also produce salivation in spite of the fact that they were never reinforced. The value of the conditioned reflex produced by these similar stimuli depends, however, on the degree of similarity to the original tone. The same can be said about stimuli of other sensory systems (i.e., other analyzers). This phenomenon was called *generalization of stimuli* (Pavlov 1960, p. 113). It is often observed both in the laboratory and in natural life.

The research on conditioned reflexes not only consisted of compiling more and more experimental material supporting Pavlov's concept, but also involved some new topics and experimental techniques. Studies performed in the laboratory of Bykov (1957) may serve as an example of such extension of research on conditioned reflexes. The results of these studies showed that it is possible to establish a conditioned reflex not only in relation to salivation but also to secretion of other glands of the alimentary system. It was demonstrated in a child provided with a stomach fistula that when the food reinforcement was always given after the sound of a trumpet, the stomach secretion started to flow to the sound of the trumpet before the food was given (experiments of Bogen, described by Bykov 1941).

The ability of conditioning was also observed on the gall bladder; namely, the contractions of the gall bladder increased when food was shown to a hungry dog (experiments of Kurtsin and Gorshkova, described by Bykov 1957).

The extension of the research on conditioned reflexes also included some new methods. An example of such novelty were experiments conducted by W.H.Gantt who also was one of Pavlov's pupils. Gantt chose the functions of the heart and the circulatory system as the object of research and demonstrated that the rules of conditioning also extended to this area (Gantt 1944, 1960, Gantt and Hoffman 1940).

All the above described experimental results, which represent only a small part of the pioneer achievements of Pavlov and his pupils, were obtained with the use of *single intermittent stimuli*. In their research there was no problem of the situational background. However, Pavlov was aware of the existence of a great number of various stimuli in the environment which could influence the experimental results. In order to avoid interference of external stimuli during experiments with conditioned reflexes, a special building (known also as a *"tower of silence"*) was constructed (see fig. 3 at the beginning of Lecture 3, Pavlov 1960).

This building consisted of four chambers on each of three floors. Each chamber was fully isolated from the external environment, securing stable conditions for the experiment. This allowed the investigator to concentrate on the effects of the intermittent stimuli and not pay attention to the situational background against which these stimuli were applied.

This led to an almost complete neglect of the problem of the situation in Pavlov's laboratory and later in the laboratories of Pavlov's followers. In fact, by using special training methods, the problem of the environmental background was removed from most experimental studies on conditioning.

Nevertheless, the problem of the situation exists. It will be discussed in further chapters of this book.

2

Instrumental Conditioned Reflexes

As usually happens in science, new theories develop by the addition of previously unknown facts or by different interpretations of the known facts. Although Pavlov's concept and theories of the conditioned reflexes seemed to encompass the whole problem of behavior, one of its mechanisms still remained unclear. That was the mechanism of the motor conditioned reflexes.

According to Pavlov (1951), the general rule of the conditioned reflex also included the motor conditioned reflexes. When a movement is performed, it produces a sensory input of impulses from the muscle. When this input is each time followed by the reinforcement (e.g., food), the conditioned salivation appears to the movement (such as a flexion of the dog's leg) before the food has been offered. This was demonstrated by Krasnogorski (1911; cit. by Pavlov 1951), who used a special device to produce a *passive* lifting of the dog's leg. In this case, therefore, the motor conditioned stimulus was indeed analogical to the acoustic, visual, or tactile conditioned stimuli (see a scheme of this conditioned reflex in fig. 2.1).

An important question remained unanswered. In order to get the motor sensory input, the movement must be performed first. In the laboratory it was produced through the use of a special device. But how is the movement initiated and performed in natural conditions? This question was not answered by Pavlov's concept (see Windholz and Wyrwicka 1996).

The problem of the motor conditioned reflexes was experimentally analyzed by two medical students of Warsaw (Poland) University, Jerzy Konorski and Stefan Miller. After reading Pavlov's

FIGURE 2.1

MOVEMENT   ⟶   IMPULSES   ⟶   SALIVATION
(classical CR)

A diagram of motor conditioned reflex in experiments of Krasnogortski according to the explanation of I.P. Pavlov (see Ch. 2).

Movement, a passive movement of the dog, evoked by a special devise (provided by the experimenter).

Impulses, nervous impulses deriving from the performed movement to the brain sensory system.

Salivation, conditioned reflex of salivation.

As shown on the diagram, the conditioned reflex of salivation appears only *after* the performance of the movement by the dog. In the laboratory, the movement appears due to the use of a technique producing the passive movement. However, there is no explanation of origin of the active movement of the animal in free conditions (without provoking any passive motor activity).

just published book on conditioned reflexes, they became enthusiasts of Pavlov's concept of conditioned reflexes. They found, however, that there was an area in the behavior of the organism that was not included in Pavlov's concept. This area was the conditioned motor behavior (occasionally also called by them "voluntary motor behavior"). Konorski and Miller, therefore, started their own experiments in the provisional laboratory space offered to them by a friendly professor of psychology.

The design of their first study was very simple. Using a mechanical device which could be controlled by them from a distance, they produced a flexion of the dog's leg to an acoustic stimulus which was immediately followed by food. After several repetitions of this procedure, the dog started to perform the leg flexion actively (i.e., without any help of the mechanical device) to the acoustic stimulus. Using various kinds of reinforcement (positive reinforcement such as food, or negative reinforcement as a puff of air to the ear, as well as the presence of the reinforcement in some cases and its absence in other cases), they found that four varieties of the motor conditioned reflexes exist. Let us briefly describe each of these varieties.

First variety. When a specific movement (e.g., flexing the leg) is performed to a conditioned stimulus (e.g., a tone) and is always followed by obtaining a reward (e.g., food), then this movement will always spontaneously appear to this stimulus.

Second variety. When a specific movement (e.g., leg flexion) appearing to a conditioned stimulus (e.g., a tone) is never rewarded (e.g., by food), whereas the reward is offered only to the conditioned stimulus alone (without the movement), then the leg's flexion stops being performed, and in its place an antagonistic movement (usually an extension of the leg) will appear to the conditioned stimulus.

Third variety. When a specific movement (e.g., leg flexion) appearing to a conditioned stimulus (e.g., a tone), is always rewarded by withholding a punishing stimulus (e.g., an airpuff to the ear), then the specific movement will always appear to the conditioned stimulus. This variety is often referred to as an "avoidance reaction."

Fourth variety. When a specific movement (e.g., leg flexion) appearing to the conditioned stimulus (e.g., a tone), is always followed by a punishing unconditioned stimulus (e.g., an air puff to the ear), then this movement will cease to appear and an antagonistic movement (e.g., leg extension) will be performed instead.

Each variety was experimentally documented. The young investigators, however, needed to hear an opinion about their research and discuss the results. Therefore, they wrote to Pavlov informing him about their study. In answer, Pavlov invited Konorski and Miller to his laboratory in Leningrad, where they had an opportunity to repeat and confirm their previous experimental results on a greater number of dogs and under better experimental conditions (Miller and Konorski 1928, Konorski and Miller 1936, Konorski 1937, 1948, 1967).

Konorski and Miller concluded that the motor conditioned reflexes differed from the Pavlovian reflexes in that the reinforcement in the former was dependent on the performance (or inhibition) of a specific movement, whereas there is no such condition in the Pavlovian reflexes. Therefore, they called the motor conditioned reflexes "conditioned reflexes type II," while reserving the name "conditioned reflexes type I" for Pavlovian reflexes. Later the terms "classical conditioning" for conditioned reflexes type I, and "instrumental conditioning" for conditioned reflexes type II, were introduced by Hilgard and Marquis (1940, Kimble 1961) and became commonly used by the behavioral scientists. Konorski and Miller admitted that the prototype of the conditioned reflexes type II was studied earlier by Thorndike (1911) who called this kind of behavior "the trial-and-error learning."

In 1948, Konorski published a book entitled *Conditioned reflexes and neuron organization* (dedicated to I.P. Pavlov and C.S. Sherrington), in which he tried to explain the experimental results obtained in Pavlov's laboratory according to the newest findings of neurophysiology, based mostly on Sherrington's studies of the spinal cord (Sherrington 1929, 1947). Konorski's new interpretation included a theoretical explanation of the phenomena called by Pavlov "irradiation," "negative induction," "positive induction," "internal inhibition," and others. Instead of the explanation of the formation of the conditioned reflex by "the meeting of the waves irradiated from different points" (of the cerebral cortex) Konorski proposed "the formation of excitatory synaptic connections between two coupled center...." (Konorski 1948, p. 251). He also proposed that the "chief feature of the conditioned reflex" is not the excitation of the conditioned centre but, instead, the excitation of the "unconditioned centre" *(p. 251)*. He also interpreted the "negative induction" as "inhibition of the conditioned reflex by an antagonistic reflex" (p.251) and the "positive induction" as "summation of the excitato-inhibitory conditioned reflexes with facilitation predominant" (p.253). The process of "internal inhibition" which was interpreted by Pavlov as an "inhibitory process occurring in the conditioned centre," was explained by Konorski as "Inhibition and excitation of unconditioned centre" or "Pure inhibition of unconditioned centre" (p.252). The full list of Konorski's new interpretations related to the conditioned reflexes can be found in pages 251–254 of his book (Konorski 1948).

Early observations of Konorski and Miller on conditioned reflexes type II (Miller and Konorski 1928, Konorski and Miller

FIGURE 2.2

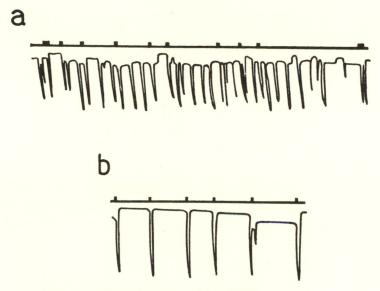

ONE OF THE FIRST RECORDS OF TYPE II CRs

Upper graph: CS (lamp).
Lower graph: record of the movements of the left foreleg.
a) The movement is performed in the presence of the CS and in intervals.
b) The movement is performed only in response to the CS.
(Konorski and Miller, 1933).

A copy of an original recording of the dog's movements (lifting of the left foreleg).

a) The dog performs the movement not only during the action of the conditioned stimulus (CS) but also during the intervals between the successive applications of the CS.

b) When food is offered exclusively in the presence of the CS, and never in its absence, the movements appear only during the action of the CS.

1936, Konorski 1967) included the fact that the trained instrumental movement appears not only in the presence of the conditioned stimulus but also in its absence, i.e., in response to the experimental situation. As shown in part a of fig. 2.2 (which is a copy of an original record taken during the early period of the training), the instrumental movement (lifting the left foreleg) appears not only during each application of the conditioned stimulus but also in the intervals between the successive applications of CS. When the re-

FIGURE 2.3

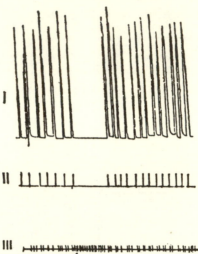

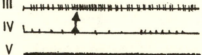

ALIMENTARY TYPE II CR ESTABLISHED
TO THE EXPERIMENTAL SITUATION

From top to bottom: movements of the foreleg, pressing the lever, salivation.
Arrow denotes the increased portion of food.
(Konorski and Miller, 1933).

The establishment of an instrumental reflex to the situation. When in the absence of the intermittent stimulus each lifting of the right foreleg and placing it on a platform was followed by food, this movement became conditioned to the situation.

inforcement (food) was given exclusively in the presence of the CS, the instrumental movements started to be performed only during the presence of CS and not in its absence (fig. 2.2, part b).

In another experiment, the dog was trained to lift its right foreleg and place it on a small board; each movement was rewarded with food. No intermittent CS was used, therefore this instrumental movement was performed to the *situation* only (fig. 2.3). When food reinforcement ceased to follow each movement, the trained movement gradually disappeared. This supported the view that the instrumental conditioned reaction was established to the situation, and not to a specific single stimulus.

These documented observations are of great importance for the theory of conditioning. They show clearly that *the conditioned movement is connected not only with the intermittent CS but also with the situational background against which it appears.* The movements appearing in the intervals between the applications of the CS are usually extinguished by being nonreinforced. Nevertheless, later observations showed that the intertrial instrumental movements can still occasionally reappear (personal observations of the author).

Early studies of Konorski and Miller (Miller and Konorski 1928, Konorski and Miller 1936, Konorski 1939, 1967) included the phenomenon of "switching" or "interchange" which consisted of the appearance of different reactions to the same CS in specific circumstances. This problem will be discussed in the following chapter.

Konorski's experimental studies on conditioned reflexes conducted in the later period comprise the problem of "plasticity" of the conditioned reflexes, that is, the possibility of the transformation of conditioned reflexes type I into conditioned reflexes type II (Konorski and Wyrwicka 1950), an exchange of excitatory into inhibitory conditioned reflexes (Konorski and Szwejkowska 1952a,b) or alimentary into defensive conditioned reflexes, and vice versa (Konorski and Szwejkowska 1956).

All the studies mentioned above, as well as most studies conducted in his laboratory by his pupils and associates, were described and analyzed by Konorski in his last book entitled *Integrative Activity of the Brain. An interdisciplinary approach* published, in 1967.

Motor behavior was the subject of studies by several investigators with a psychological rather than physiological approach. In 1911, Thorndike published his book *Animal Intelligence,* in which he described his experiments on learning of various motor tasks by animals such as fish, chicks, cats, dogs, and monkeys. Fish had to find an opening in a partition inside the aquarium in order to reach the food placed behind the partition. Chickens had to find the correct way in a maze to the food placed at the goal. The required motor tasks for cat and dogs included pressing a lever, pulling a cord, turning a knob, pushing the door, or in some special experiments, licking the fur or scratching. Monkeys had to perform more complex movements, such as manipulations with a wooden peg, hook, metal bar, or turning a knob by 270°, to get a reward. This kind of behavior, which was called by Thorndike "trial-and-error learning," corresponds to Konorski and Miller's "conditioned reflexes type II" (as mentioned earlier).

Extensive studies on the problem of motor conditioned behavior were conducted by Skinner (1938, Ferster and Skinner 1957). Skinner also distinguished two types of behavior: the type corresponding to classical conditioned reflexes was called the "respondent behavior," and the type corresponding to conditioned reflexes type II of Konorski and Miller was called the "operant behavior." The respondent type of behavior refers "to conditioning which results from the contingency of a reinforcing stimulus upon a *stimulus*" whereas the operant type of behavior results "from contingency upon a *response*" (Skinner 1937, pp.272–73).

Skinner developed special techniques for studying the operant behavior. His subjects were either pigeons or rats. The operant task for pigeons was to peck at the visual stimulus and for rats to press the lever. The movements of the subjects were automatically recorded enabling an immediate analysis of the obtained data. This resulted in the compiling of a great amount of experimental data related to such problems as discrimination of responses, generalization of stimuli, delayed responses, inhibition (extinction), chained schedules of conditioning, and others (Ferster and Skinner 1957). Skinner also theorized on problems such as "drive" and "drive and conditioning," and discussed the effects of emotions and drugs on conditioning (Skinner 1938). However, he did not relate the behavioral results of his studies to brain functions.

One of the most important points of Skinner's theoretical approach was the nature of the stimulus. Skinner emphasized the fact of the presence of great variability of stimuli in the environment and the necessity of the differentiation of the chosen stimulus before its conditioning. He also stressed that a chosen motor act to be conditioned is never completely new but it could be performed occasionally earlier. Therefore, when speaking about the conditioned motor behavior we should use a comparative measure: to prove that the number of performances of this motor behavior is significantly greater after than before conditioning.

Because all motor activity of the organism is based on the inborn motor reflexes, it is possible that the movement chosen for the training is not new but could have been performed occasionally in the history of the subject. Being aware of this, the investigator must be sure that the chosen movement was never performed by the subject during the preliminary control sessions. The appearance of that movement during the training sessions can be considered proof of the establishment of this movement as an instrumental conditioned reflex. The problem of the *absolute* novelty of the movement should never be considered.

The above discussed points are only a part of Skinner's hypotheses and suggestions. The interested reader is directed to Skinner's original publications (Skinner 1938, Ferster and Skinner 1957).

3

The Phenomenon of Switching

As mentioned previously in chapter 2, early research on instrumental conditioning has shown that, in some circumstances, a definite conditioned stimulus may not produce a motor reaction trained to it, but instead, a different reaction that has been trained to a different stimulus before (Miller and Konorski 1928; Konorski and Miller 1933, 1936; Konorski 1939, 1948, 1967). Here are some examples of their observations.

In studies on this phenomenon, called by Konorski and Miller (1928, 1936; Konorski 1939, 1967) cases of "switching" (also called by them "transfer" or "interchange"), the authors tried to find the exact conditions which determined the appearance of this behavior.

Switching in Homogeneous Instrumental Reactions with Different Defensive Reinforcement

In a dog, a classical (Pavlovian) conditioned reflex (CRI) had been established when a tone was always reinforced with the introduction of a small portion of acid solution into the dog's mouth. The conditioned reaction consisted of salivation and slight defensive movements. In further sessions, the sound of a bell was always followed by a puff of air into the dog's ear; however, when the dog's left foreleg had been passively lifted during the action of the bell, the puff of air was not given; as a result, an instrumental avoidance reaction (i.e., active lifting of this leg) was established. When this reaction was firmly established, the previously used CS, the tone, was applied. Already after several trials, the tone began to evoke both salivation (the CRI previously trained to the tone) and the lifting of the left foreleg (CRII, the reaction trained to the bell but never to the tone). According to Konorski's interpretation, the obtained result demonstrated that a CS connected by training with a definite defensive CRII, may elicit this reaction to another CS related to a different defensive reinforcement. Konorski also hypothesized that similar cases of switching may also occur in conditioned reflexes with approach type of reinforcement (Konorski 1939, p. 20). This supposition was verified by Fonberg (1961, 1967), as described later in this chapter.

Switching in Heterogeneous Instrumental Reactions

This type of experiment was performed by Konorski and Miller during their visit to Pavlov's laboratory (Konorski and Miller 1936, Konorski 1939). First, an alimentary instrumental reaction of lifting the right hindleg to a specific noise and a defensive (avoidance) reaction of lifting the right foreleg to the sound of bubbling water were established in a dog. During each session a band was fastened on the trained leg for recording purposes. It was found that when the band was fastened on one leg only, the dog lifted this leg both to alimentary and defensive stimuli. Only when the bands were fastened on both trained legs, were the reactions correct, i.e., the alimentary CS produced the alimentary CRII, and the defensive CS produced the defensive (avoidance) CRII. However, when the alimentary CS was applied after a number of trials with the defensive CS, the dog first performed a defensive CRII and then only the correct alimentary CRII. And, vice versa, when the defensive CS was preceded by a number of successive applications of alimentary CS, the alimentary CRII appeared first and then only the correct defensive CRII was performed. This case of switching (regarded by Konorski and Miller as a result of the "inertia of excitation") was not always observed in all dogs.

Switching Produced by Factors Other Than Inertia of Excitation

An alimentary CRI was established separately to the metronome, to a tactile stimulus, and to the lighting of a lamp; in addition, a defensive CRII consisting of lifting the left foreleg to avoid a puff of air into the ear, was established to the sound of a bell. Then a process of extinction of the CRI to the metronome was performed. During a number of sessions the metronome was applied 2–10 times per session without the food reinforcement, whereas the tactile stimulus and the lamp were always followed by food. The band (for recording purposes) was always fastened on the left foreleg. During the first 17 extinction sessions salivation to the metronome beats was only partly decreased. When, starting with the 18th session, the metronome was applied many times in a row in each session without food reinforcement, the salivary CRI rapidly diminished and the avoidance lifting of the left foreleg (established to the bell but never used with the metronome), began to appear both during the action of the metronome and during the intertrial intervals. Thus, *a deep extinction of an alimentary CRI resulted in the performance of a defensive CRII to the alimentary CS.*

A question was raised as to whether the alimentary instrumental reaction may appear to a defensive CS in conditions different than the inertia of defensive excitation. In order to answer this question, an alimentary classical type conditioned reaction (CRI) was established in a dog to the tactile stimulus, the lamp, and beating of the metronome (120/min). In the same dog, an alimentary instrumental reaction (CRII) of lifting the right foreleg to the sound of a bell, and a defensive instrumental reaction of lifting the right hindleg to the tone to avoid an air puff to the ear was established. Then a new defensive CS, the sound of bubbling water, reinforced by acid-to-mouth, was introduced. It was found that already the very first application of the unconditioned stimulus, acid-to-mouth, produced the defensive movement of the right hindleg. In the following trial, the same movement appeared to the bubbling. In the third trial, both the lifting of the foreleg (an alimentary CRII) and the lifting of the right hindleg (a defensive CRII) were performed to the bubbling of water. A similar effect was observed when a new stimulus, the sound of a rattle, was followed by a puff of air. Already in the third trial the rattle evoked lifting the right hindleg (a defensive CRII, previously established to the tone) and then lifting the foreleg (an alimentary CRII, previously established to the bell). These experiments showed that *the alimentary CRII may appear to a defensive CS in conditions others than inertia of preceding excitatory processes.*

The obtained results were interpreted by Konorski and Miller (1933, 1936; Konorski 1939, 1967) in the following way. There are two kinds of stimuli: the *determining* and the *releasing* stimuli. The determining stimuli, such as a band on the dog's leg, or overall experimental situation, do not elicit the conditioned reaction but they determine *which kind* of reaction is to be performed; the alimentary or defensive background produced by preceding application of unconditioned stimulus could also be considered a determining stimulus.

The releasing stimuli are usual intermittent stimuli that *elicit* the reaction. As previously described, a partly inhibitory (differential) alimentary CRI could elicit a defensive CRII trained to other CS; also, a deep extinction of an alimentary CRI could result in the appearance of a defensive CRII established to a different CS. An application of an aversive unconditioned stimulus could also release a previously established alimentary CRII.

The boundary between the determining and the releasing stimuli is not absolute, however; for instance, a continuous determining stimulus may be, at the same time, a releasing stimulus, and vice versa, a sporadic stimulus may both determine and release the reaction (Konorski 1967, p. 389).

As already mentioned earlier in this chapter, Konorski and Miller observed a transfer between two CRIIs, each of which was trained with a different aversive unconditioned stimulus.

Later studies by Fonberg (1961, 1969) showed that the transfer of instrumental reactions is possible both in avoidance and in alimentary CRIIs, as described below.

Transfer in Avoidance Behavior

Two different avoidance CRIIs were trained in four dogs in the usual chamber. The conditioned stimuli were the whistle, the bell, the tone, and the light. The training was conducted in separate series to each CS. Each animal was trained to lift the right hindleg to the CS1 to avoid a nociceptive reinforcing stimulus which was either an electric shock or a puff-to-ear in various dogs. Then, in another series of sessions, the dog was trained to place its left foreleg on a platform to stimulus CS2 to avoid a nociceptive stimulus. After the establishment of the second CRII, a test was performed in which CS1 and CS2 were applied in the same session. It was found that both reflexes (lifting the right hindleg to CS1 and placing the left foreleg on the platform to avoid nociceptive stimulus) were usually performed correctly and they did not interfere with each other when the animal was calm. However, when the animal was excited, both avoidance reactions appeared in the same trial (to either CS1 or CS2), often simultaneously or almost simultaneously; these movements also were observed in the intertrial intervals. This kind of switching was observed in two dogs.

According to Fonberg's interpretation, these results support a hypothesis that there exist specific centers in the brain, corresponding to each nociceptive stimulus; these centers are all connected to a common defensive center. In the state of excitement (such as fear) the common center may activate the specific defense centers, resulting in the appearance of all trained avoidance responses to each CS.

Approach Behavior

In another study (Fonberg 1967), two dogs were trained to press a lever to the tone to obtain food. Then chronic electrodes were implanted in these dogs' lateral hypothalamic "feeding" area (Anand and Brobeck 1951). Electrical stimulation of the hypothalamic points corresponding to the tips of the electrodes, produced eating in satiated animals. When the hypothalamic stimulation was substituted for food reinforcement, the dogs began to perform the previously trained alimentary CRII (lever pressing).

In further sessions, the instrumental response was reinforced either with food or with the electrical stimulation in the lateral hypothalamus. It was observed that the hypothalamic stimulus did not produce an increase of hunger but rather a state equivalent to food reward. In this experiment, therefore, a specific kind of switching took place between two different reinforcements of the approach type, the food and the electrical stimulation in the lateral hypothalamus. According to Fonberg's interpretation, the hypothalamic stimulation could produce sensations similar to those provided by food intake, leading to the performance of the response that was originally trained for food reinforcement.

The Studies on Switching Conducted in Asratyan's Laboratory

The phenomena of switching were extensively studied on dogs in Asratyan's laboratory with the use of the classical conditioned reflexes (Asratyan 1951, 1961, 1965). In the first study, in sessions conducted in the morning by Shitov, the CS of 120/min beats of the metronome was always reinforced with food. In the same dogs, in sessions conducted in the afternoon by Yakovleva, the same stimulus (metronome beats of 120/min) was reinforced by an electric shock. As a result, in the morning session the beats of metronome elicited an alimentary reaction (salivation and approach to the feeder), and in the afternoon session the same CS produced a defensive reaction of lifting the leg that was to receive the electric shock.

Even more complex experiments on switching were conducted by Struchkov (Asratyan 1961). First, two conditioned reflexes were established in each dog, an alimentary CR to the sound of a buzzer (reinforced with food) and a defensive reflex to the tactile skin stimulation reinforced with an electric shock to the skin. These reflexes were trained in the same experimental compartment. Then, a parallel series of sessions began in another experimental compartment in which the previous CSs were used but the reinforcements were reversed. That is, the sound of buzzer was now reinforced with electric shock, and the tactile stimulus was reinforced with food. After some training each of these CSs could evoke two different conditioned reactions, an alimentary CR in one compartment and a defensive CR in the other compartment.

In experiments conducted by Pressman (Asratyan 1951) stimulating electrodes with a registering device were attached to the dog's left hindleg and a defensive conditioned reflex was established to the tactile stimulation of that leg. Then the stimulating electrodes and the registering device were attached to the dog's right hindleg (which was never used in experiments before), and the usual experimental procedure started. This time, the CS evoked lifting the right hindleg and not the left hindleg which was previously used in the training. When the electrodes were attached to both hindlegs, the CS evoked lifting either the left or right hindleg, in irregular manner. However, after application of an electric shock to one of these legs, the dog began to lift only that particular leg in further trials. These experiments are similar to those performed by Konorski and Miller who obtained the same results when the band was fastened on both hindlegs of the dog.

In other experiments performed by Shitov and Zamyatina (Asratyan 1961, 1965), each experimenter conducted two sessions daily, one in the morning and another in the afternoon. In the morning session, the acoustic CSs were reinforced with food, and in the afternoon they were not reinforced at all. As a result, *the same acoustic CS elicited an alimentary (salivary) CR in the morning session and an inhibitory (no salivation) CR in the afternoon session*. This experiment showed that switching may occur not only between two excitatory reactions but also between the excitatory and inhibitory states.

According to Asratyan (1965), the phenomena of switching are a result of action of permanent situational stimuli called by him *tonic stimuli*. In some cases, the tonic stimulus was a different experimenter, and, in other cases, a different time of the day. A different kind of reinforcement in the afternoon session than that in the morning session could also be a factor in switching, playing the role of a tonic stimulus. The tonic stimuli do not evoke the conditioned reaction but they may facilitate the action of *phasic stimuli*, i.e., the intermittent stimuli (such as a tone).

The effects of tonic stimuli were demonstrated by Sakhiulina (1955, cit. by Asratyan 1965, pp. 139–140) who recorded the electroencephalographic responses (EEG) from various points of the dog's cerebral cortex during the experiments on switching. In the morning session, an auditory stimulus was paired with electric shock to the dog's left hindleg; in the afternoon session, the same auditory stimulus was paired with an electric shock to the right hindleg. This training resulted in lifting the *left hindleg* to the CS in the morning session and lifting the *right hindleg* to *the same CS* in the afternoon session. At the same time, the EEG record showed the presence of high amplitude waves in the leads from the boundary area between the motor and parietal areas, ipsilateral to the trained legs, i.e., from the left hemisphere in the morning sessions and from the right hemisphere in the afternoon sessions. This EEG pattern appeared at the beginning of the session, before the application of the CS, and persisted throughout the session. This kind of EEG activity could reflect the action of the tonic stimulus.

Other Studies on Switching

More recently, the problem of switching was studied on human subjects by a number of investigators (Kimmel and Ray 1978, Murrin and Kimmel 1986, Lachnit 1986, Kimmel and Lachnit 1990, and others). In their studies, a strictly defined stimulus was selected, often referred to as a "context" or "contextual stimulus," or, according to the terminology used by Asratyan, a "tonic" stimulus against which a "phasic" (i.e., intermittent) CS was acting. In their experiments, the tonic stimulus preceded and lasted through the end of duration of the phasic stimulus. The reinforcement was an electrodermal shock to the finger. In the case when the tonic stimulus preceded CS, the reinforcement was withheld. The results of these experiments showed that in cases of the application of the tonic stimulus the conditioned response was absent. This effect was considered by the experimenters a case of switching.

These experiments were similar to studies of Pavlov's laboratory on "conditioned inhibition" in which the reinforcement was withheld each time a novel stimulus preceded the action of the CS (Pavlov 1960, Ch.5). The difference between the "context" experiments and Pavlovian studies was that the context studies were conducted on human subjects whereas Pavlovian studies were conducted on dogs.

Studies on the effect of the "context" were also conducted on pigeons by Rescorla (1984). In his experiments the contextual stimulus was the design of lining of the walls of the experimental compartment combined with a visual stimulus (blue or yellow light); some of these combinations were reinforced, whereas other were not reinforced. The results of these experiments led the author to conclude that CSs become associated with the context in which they occur, and that manipulations of CS (by reinforcing or not reinforcing it) produce changes in the associations between the context and the CS which, in turn, influence the performance to the other CSs used in these experiments. An especially striking result was that related to the extinction experiment. It showed that the extinction process related to one CS strongly influenced the subsequent behavior related to the other CS.

The experimental data described in this chapter are strictly related to the influence of the situational background on behavior. They also support the hypothesis that conditioning of the situation does take place, and its effect on behavior dominates over the effect of the single conditioned stimulus. This problem will be further discussed later in this book.

4

Cases of Dependence of Conditioned Behavior on Situational Factors

Observations similar to those reported by Konorski and Miller (1933, 1936; Konorski 1939, 1967) were made in the laboratory of Kupalov (Denisov and Kupalov 1933, Kupalov 1961). In a study on dogs, two kinds of illumination of the experimental chamber were used, full illumination in some sessions and poor illumination in other sessions. It was observed that when the conditioned reflex was trained in full light and then tested in poor light, the value of the reflex remained at a high level for several following sessions, and then only decreased and stabilized at a low level. And, vice versa, when after several weeks of training in poor light full light was used again, the conditioned reflex remained at low value for the next several days.

Kupalov (1961) explained this phenomenon as an effect of conditioning of the experimental situation. According to him, there are two kinds of brain processes involved in conditioning. One process is related to the intermittent conditioned stimulus that produces the conditioned reaction. The other process is related to the experimental situation (such as the kind of illumination) which does not produce the reaction but may change the *tonus* of certain parts of the brain. Kupalov called this second process a "*shortened*" or "*truncated*" conditioned reflex. According to Kupalov, therefore, the role of the experimental situation is limited: it can only modify the tonus of the excitatory process but cannot play a decisive role in the performance of the reflex. Kupalov stressed, however, that the strength of these "shortened"

31

reflexes can be unexpectedly high. This interpretation will be discussed in chapter 9, together with other theoretical problems.

A specific case of situational influence on conditioned reflexes was observed by Doty and Giurgea (1961; Giurgea 1989). In a dog, they paired moderate electrical stimulation of the *left* occipital cortex with similar stimulation of the *left* motor cortex. As a result, stimulation of the left occipital cortex regularly produced a flexion of the right forepaw and turning of the head to the right (as if they stimulated the left motor cortex). When conditioning was well established, electrical stimulation of the *right* homotypical cortical regions including the right ectosylvian area was applied. It turned out that this time, stimulation of either the right occipital cortex or the right motor cortex produced regularly the previous response, i.e., a flexion of the right forepaw and turning head to the right. Therefore, the motor reaction remained the same, as previously trained, no matter which, left or right, cortical area was stimulated. The authors called this phenomenon a "conditioned transfer."

However, this transfer was observed only when the general experimental situation was unchanged. After placing the animal on the experimental table and then rotating the table by 180°, the previously observed effect was no longer observed. That is, stimulation of the left occipital cortex or left motor cortex produced normal contralateral movements, whereas stimulation of the right occipital cortex remained ineffective. On the other hand, stimulation of the right motor cortex elicited contralateral reactions, i.e., head rotation to the left and flexion of the left forepaw. As commented by Giurgea (1989) "even such as ancient, phylogenetically and ontogenetically rooted physiological response as motor contralaterality following efficient motor cortex stimulation might be critically modulated by shortened conditional reflex in the experimental situation."

The influence of the experimental situation in conditioning was earlier demonstrated in studies conducted in Konorski's laboratory. The topic of the first of these studies (Konorski and Szwejkowska 1950) was the chronic extinction of conditioned reflexes against an *excitatory background*. In two dogs, several alimentary classical conditioned reflexes were established to various stimuli. One of these CSs was chosen for chronic extinction. This stimulus was applied, without being reinforced with food, only once during each session, among other conditioned stimuli which were reinforced. After a number of sessions, the conditioned salivation stopped appearing to the non-reinforced stimulus which became an inhibitory stimulus. Then the extinguished conditioned reflex to this stimulus was restored by offering food again after each application of this stimulus. The same procedure was then used with each of the remaining stimuli. It appeared that chronic extinction was dependent on the strength of each stimulus and a degree of its fixation; the acoustic CSs (which produced strong salivation) required 21 to 62 sessions to become extinguished whereas the visual CS (which produced a weak salivation) required only 15 sessions.

In a further series of experiments, the chronic extinction against an *inhibitory background* was studied on two other dogs by Szwejkowska (1950). This time the conditioned stimulus used for extinction was applied in separate experimental sessions during which the food was not given at all. Such sessions were interspersed with sessions in which the other stimuli were applied with the usual reinforcement. It was found that the chronic extinction of a conditioned reflex against an inhibitory background occurred *more rapidly* than the chronic extinction against the excitatory background, requiring only 16 sessions in the case of using a strong acoustic stimulus, a bell.

The results of both studies (Konorski and Szwejkowska 1950, Szwejkowska 1950) suggest that the differences in the results were caused by the use of two different situations in the process of extinction. This supports the hypothesis that the experimental situation plays an important role in the conditioning processes.

The role of the situation in the extinction of the instrumental conditioned reflexes was also studied by Wyrwicka (1952). In these experiments, conducted on three dogs, the course of extinction was studied against either an excitatory or inhibitory background. In the experiments on the excitatory background, only one conditioned stimulus, the sound of bubbling water, was used. This stimulus was applied several times in each experimental sessions at intervals of 3–5 minutes and was not reinforced with food. However, food was given during the intervals, without any relation to the intermittent stimulus. Already on the first session of this study, the latency of the instrumental reaction to the CS, which was 4 seconds in the first trial, increased to 12 seconds in the 7th trial. On the other hand, the salivary CR which was 35 points on the manometer scale, in the first trial, decreased to 18 in the 7th trial although the salivation during eating remained about the same up to the end of the session. During the following sessions both salivary and instrumental conditioned reactions to the CS gradually diminished and after 6–8 sessions they completely disappeared. However, as soon as the reinforcing stimulus, food, was given again to the extinguished stimulus (bubbling), both salivary and instrumental conditioned reflexes gradually recovered.

Several months later, another extinction series of sessions was conducted on the same dogs. This time, however, a different conditioned stimulus, a whistle, was used, and no food was offered either to CS or during the intervals between the successive trials. This was, therefore, an extinction against the *inhibitory background.* When the reflex to the whistle was entirely extinguished, the reflexes to other stimuli were tested. It appeared that both instrumental and salivary reactions to these stimuli (which were not used during the process of extinction) were *strongly diminished.* Similarly to the extinction against the excitatory background, both salivary and instrumental CRs gradually recovered as soon as food reinforcement was offered again.

The changes in the experimental procedure also appeared an important situational factor in behavior. In experiments on dogs by Zbrozyna (1953) an acoustic or visual (light) stimulus was applied when the dog was eating and a few seconds later the food was withdrawn. After several repetitions of this procedure, the dog stopped eating and turned away from the feeder as soon as the withdrawal signal was given. However, when the same stimulus was applied before food was offered it evoked salivation and approach to the feeder. Therefore, the same stimulus could elicit two different reactions depending on the procedural situation, in the presence or absence of food. In the presence of food the conditioned stimulus evoked a negative reaction, i.e., cessation of eating and turning away from the feeder; and in the absence of food, the same stimulus produced a positive feeding reaction. The author interpreted these results as a phenomenon of *non-identification* of a stimulus appearing against a different situational background.

Cases of Dependence of Behavior on the Experimental Situation in Psychopharmacological Studies

In a study of Giurgea (1970) on dogs, two classic conditioned reflexes were established in each animal. One of these reflexes was trained in the usual Pavlovian room, and another reflex was trained in Kupalov's "free-behavior" room, to the same conditioned stimulus. The two rooms were separated by a long (of about 30 m) corridor. In the middle of the corridor there was a small room in which injections were given. It was found that chlorpromazine (1 mg/kg, intravenously) strongly reduced the value of conditioned reaction when the dog was tested in the Pavlovian room 10 minutes later.

On the other hand, when the dog was tested on another day, in the free-behavior room, the same amount of chlorpromazine did not have any effect on conditioned reaction. The same dog, after injection of this drug, taken just in front of the Pavlovian room, showed a full neuroleptic effect, whereas when placed in front of the free-behavior room, this dog appeared almost normal as if he did not receive any drug. When taken back in front of the Pavlovian room, this dog again showed the neuroleptic behavior whereas no such behavior was observed when placed in front of the free-behavior room. Such alternation in behavior could be induced 10–12 times during a period of 30–45 minutes, in various dogs.

This finding supports earlier observations of Kupalov (1961) that dogs were more resistant to neurogenic procedures in the free-behavior situation than in the Pavlovian room. The situational effects on behavior were also demonstrated in studies of Zuckermann and Buffy (1960), and more recently by Khananashvili (1983).

Poulos and Hinson (1982), in a study on rats, also observed a

strong dependence of the effects of the neuroleptic drug haloperidol on the environmental background. They found that about
20 daily injections of this drug resulted in the development of a
highly significant tolerance to the cataleptic effects. Then, a group
of rats was injected twice daily, once in room A with haloperidol, and once in room B with saline, in random alternation from
day to day (but always with haloperidol in room A and saline in
room B). After about 20 days, tolerance to haloperidol was present
only in room A where the rats received haloperidol. No tolerance was present after haloperidol injections in room B (where
the rats usually received saline); instead, an almost total catalepsy was seen.

In studies of Corson (1967), and Corson and O'Leary-Corson
(1968), dogs with salivary fistula and urinary bladder fistula were
tested for electroencephalic and respiratory activity in two different rooms. One of these rooms was an experimental compartment in which the dogs received painful shock to a hindpaw to a
tone. Another room was a control compartment in which the dogs
never got a painful shock. When tested in the control room during a 3–hour session, the dogs urinated a normal amount of low-
density urine; they also showed no spontaneous salivation and
their heart rate lowered over time. However, when tested in the
experimental room, where they obtained shocks following a tone,
the dogs urinated much less with a high density urine, produced
some salivation, and showed elevated heart rate. When they were
tested again in the experimental room, without receiving any tones
or shocks, the dogs still showed the previously observed reactions when the tones and shocks were used. This suggested that
the experimental compartment alone was able to signalize,
through a "shortened" conditioned reflex of the situation, a "dan-
ger" related to the previous experience in the experimental room
(cit. by Giurgea 1989).

In another study, conducted on morphine-reacting rats,
Eikelboom and Stewart (1979) demonstrated that saline may evoke
responses similar to the hyperthermic response usually produced
by morphine, but only when the saline injection was given in the
environment in which the rats usually obtained morphine. No hy-

pothermic response was seen in the rats' home cage where they usually did not receive morphine injections.

Eikelboom and Stewart (1979) also observed the environmental effects of daily injections of morphine on body temperature in 30 male rats. The animals were divided into four groups. one group received injections of 5 mg, the second group 25 mg and the third group an increasing dose of up to 200 mg of morphine. The fourth group received only saline injections. In each rat the rectal temperature was measured five times a day in three different environments: the home cage (a neutral environment), the pre-injection environment where the rats were placed before the daily injection, and the injection environment where the animals remained after the injection. The injections were followed by a period of abstinence.

The tests for conditioning showed that rats in the morphine groups, compared to the saline group, showed an anticipatory hypothermia in the pre-injection environment; this was just opposite of the unconditioned hyperthermia to morphine. On the other hand, the rats in the morphine groups showed a conditioned hyperthermia when tested in the injection environment after a period of abstinence. The results indicate that the changes of body temperature depended on the environment which was related to the morphine injection. In other words, it can be said that conditioning to the environment took place.

Circling behavior produced by apomorphine in rats, and its dependence on the environment were reported by various investigators (Ungerstedt 1976, Silverman and Ho 1981). It was found that even simple placing of the rats in the cage where they previously obtained apomorphine injections resulted in the appearance of circling. This effect was even stronger when the rats obtained saline injection before the test.

The environmental effects were demonstrated by Melchior (1988) in a study on tolerance to ethanol in male mice. It was observed that intracerebroventricular (ICV) injection of a very small amount of ethanol produced a brief but substantial hypothermic response. However, after eight ICV injections of 2 mg of ethanol at 2-hour intervals, four injections per day during two days, an environmental tolerance to the hypothermic effect of ethanol developed. The environment-dependent tolerance was also seen when these mice were tested with intraperitoneal injections of ethanol.

A study on gastric secretion conducted on cats by Wyrwicka and Garcia (1979) also showed the environmental influences. The experiments were conducted on two groups of cats. In one group of 6 cats basal gastric secretion, and in another group of 7 cats pentagastrin-induced gastric secretion, was collected during a 2-hour session, two or three times a week. There were two kinds of sessions. In one session, the cat's movements were restricted by placing the animal in a harness, and in the other session, the cat was unrestricted and could move freely in a cage. It was found that in 4 of 6 cats in the basal secretion group and in 5 of 7 cats in the pentagastrin secretion group, gastric acid output was significantly higher in sessions in harness than in sessions in cage. These differences in acid output were due to acidity rather than volume of secretion. It was hypothesized that the restriction of the movements in the harness produced a "reflex of freedom" which could not be accomplished because of confinement. This led to neural disturbances in the autonomic system, resulting in changes in gastric acid secretion.

Special observations on the situational effects on behavior of human subjects were reported by Lhermitte (1986). Although his observations were made in patients after frontal lobectomy, nevertheless they provided important material concerning the role of definite environmental situations in behavior.

Two patients, a man 51 (patient 1) and a woman 52 years of age (patient 2), were taken to a doctor's office, a lecture room, a car, an apartment, and a gift shop. It was found that the patients were excessively dependent on environmental cues. Here are the excerpts of observations made in various situations.

When in the doctor's office some medical instruments were put on the desk, the female patient immediately picked up the blood pressure gauge and measured the doctor's blood pressure. Then she took the tongue depressor and examined the doctor's throat.

Both patients were taken by the doctor to a lecture room with a buffet of food and beverages. Patient 1 helped himself to the food and the orange juice and behaved like a guest. In contrast, patient 2 behaved like a hostess. When she saw some stacks of chairs, she proceeded to set the chairs side by side. Then she offered some foods on various plates and a glass of orange juice to the doctor. She did not serve herself.

The doctor took patient 1 and his female friend to his apartment and said that it was a "museum." Patient 1 immediately started to examine the paintings as if he were in a real museum. His behavior was appropriate for any usual museum visitor. Moreover, when he saw that one painting was down on the floor, he took the hammer and nails (which had been left next to the painting), hammered a nail into the wall and hung the painting. Then the doctor and patient walked around the apartment and also entered the bedroom. When the patient saw the bed without the bedspread, he immediately started to undress and then got into bed and prepared to sleep.

The behavior of patient 2 in the bedroom was slightly different. She did not undress but lay down immediately. Then she accompanied the doctor to a table with various items used for intramuscular injections. When the doctor showed her a syringe, she took it and gave the injection to the doctor.

When discussing the patients' behavior, Lhermitte proposed defining it by the term *environmental dependency syndrome.*

All studies described in this chapter show the decisive role of the environmental situation on the behavior. From the point of view of conditioning, the behavior became associated (conditioned) to the situation due to the repetitive application of the reinforcing stimulus in this situation. This problem will be discussed from the theoretical point of view in a later chapter.

5

Studies on the Role of the Situation and the Intermittent Stimulus in Conditioning

As described in chapter 1, all Pavlov's experiments were conducted using single, intermittent stimuli appearing against a permanent and unchangeable environment. The problem of situation in which the conditioned stimulus appears was practically not discussed. However, later observations by Vatsuro (1948) revealed that changes in experimental setting could result in a decrease of conditioned reflexes in dogs and monkeys. Stroganov (1948) observed inhibition of conditioned reflexes after transferring experiments from one experimental chamber to another. Similarly, Beritov (1948) found that extraneous factors introduced into the usual situation might inhibit or completely change the established conditioned motor behavior.

A systematic study on the role of the experimental situation on the conditioned reflexes type II (instrumental conditioning) was undertaken by Wyrwicka (1956, 1958, 1993). The study consisted of several series of experimental sessions conducted daily for a two- week period, separately for each of 2–5 dogs in each series. The sessions were conducted always about the same time of the day. The reinforcement was food (bread) in all experimental series.

First Series: One Experimental Situation

In a usual Pavlovian chamber, each dog was trained to perform two reactions: one was lifting the right hindleg (CR1) to an acous-

41

tic stimulus, the sound of a rattle (CS1), and the other was placing the right forepaw on the feeder's platform (CR2) to a visual stimulus, a rotating disc (CS2). Each reaction was trained in a separate series of sessions. A two- week series of daily sessions with CS1-CR1 alternated with a fortnight series of sessions with CS2- CR2. It was observed that at the beginning of the next series each CS evoked a "wrong" reaction, i.e., the reaction trained during the preceding series (for instance CR2 to CS1, or CR1 to CS2). Only when the wrong reaction was not reinforced by food did the correct reaction appear. It was also found that during further sessions of a series, the dogs often performed both reactions to each CS (i.e., both CR1 and CR2 to either CS1 or CS2).

Second Series: Two Similar Experimental Situations

A separate group of dogs was trained in two experimental chambers, A and B. The chambers were similar to each other, but were not identical. The stimuli and reactions were the same as those used in the first series. Reaction CR1 to stimulus CS1 was trained in chamber A, and reaction CR2 to stimulus CS2 was trained in chamber 2 in fortnight series of daily sessions. The results of this experimental series were quite similar to the results obtained in the first experimental series conducted in one situation. That is, at the beginning of the new fortnight series, each stimulus evoked the reaction trained in the preceding series (CR2 instead CR1, and vice versa) and during further experimental sessions each stimulus often evoked both reactions. In addition, during an acute extinction of CRI or CR2 (by not reinforcing them with food), limited to one session only, both CR1 and CR2 appeared to each stimulus, CS1 or CS2.

Third Series: Two Different Experimental Situations

In this experimental series, each dog was trained in two situations different from each other. One of these situations was a usual Pavlovian chamber (situation A) in which the dog was trained to perform reaction CR2 (placing the right forepaw on the platform) to a visual stimulus CS2. Another situation was a large empty pen (without any stand or a platform) arranged on the floor of a room located in a separate building (situation C). In this situation the dogs were trained to perform another instrumental reaction CR3 (lifting the left foreleg in one dog, and assuming a "begging" posture in the other dog) to a stimulus called "noise."

Then, a new acoustic stimulus, the beating of a metronome, mCS, was introduced to situation A and reaction CR2 was easily established to it. After several days of training, the metronome device was transferred to situation C and stimulus mCS was applied. The reaction of one dog was the performance of reaction CR3 *instead of reaction CR2* which was recently trained to the same stimulus (mCS) in situation A. The same result was obtained when a tactile stimulus (produced by a device attached to the dog's back) was used as a new stimulus A and then tested in situation C.

The above described experiments suggest that in the process of conditioning, the instrumental reaction creates associations not only with the single intermittent stimulus but *also with the permanent complex of stimuli of the environmental background against which the stimulus is applied.* Moreover, the above experiments suggest that the role of the situation may be *dominant* over that of the intermittent stimulus.

This supposition is supported by additional facts obtained earlier (Wyrwicka 1952; also cited by Konorski 1967, p.409). Namely, it was found that when an instrumental conditioned reaction is already established to one conditioned stimulus, (e.g., a whistle), the same instrumental reaction will appear to all other new stimuli, not only acoustic, but also visual and tactile stimuli, used in the same situation (including the same experimental procedure) for the first time. This fact was explained as a result of establishment of the conditioned connections not only with the intermittent stimulus but also with the situation as a whole. When speaking about a conditioned instrumental reaction it should be remembered that it is evoked by *a complex of stimuli: the intermittent stimulus (CS) plus the situation, and not only by the single CS.*

In relation to the above described experiments, it would be interesting to quote some special experiments by Soltysik (1975). In his study, dogs were trained to perform a definite instrumental reaction to get a food reinforcement. However, when the experimental procedure was changed, namely the food reinforcement was given immediately after the start of application of a definite conditioned stimulus, the dog began to eat at once, without first performing the instrumental reaction. After training (in which the delay in offering food was gradually extended) it was possible to obtain a period up to 10 seconds of waiting for food, without the instrumental reaction. In this case, the situation included a procedure *different* from the standard procedure used with other conditioned stimuli; such a change in the experimental procedure was also a change in the situation and this resulted in obtaining a different kind of behavior. This study further supports the idea that changes in the situation may result in changes in the conditioned reaction.

Studies on changeability ("plasticity") of the conditioned stimulus by Kozak et al. (1961) and Kozak and Westerman (1966) also showed that situational factors, such as altering the temperature in some parts of the animal's body, presence or absence of food, and others, could change the reaction.

The Role of the Intermittent Stimulus

Considering the obtained data, a question may be asked— what is the role of a single stimulus in the process of instrumental conditioning? Can it retain its acquired property of evoking the instrumental reaction when it is used in an environment different than that in which it was originally trained? In order to answer this question, another study was undertaken, described as follows.

In a special study (Wyrwicka 1958, 1993), four experimental series of sessions, two of them with alimentary reinforcement ("alimentary group") and two others with defensive reinforcement ("defensive group"), were conducted on a total of 14 dogs. Three to five dogs were used for each series. The training was conducted separately for each dog of the experimental series. First, in a usual Pavlovian chamber (uE), each dog in the alimentary group was trained to lift the right foreleg and place it on the feeder platform (aCR) to stimulus aCS (the sound of a flute) for food reinforcement. In the same chamber, each dog in the defensive group was trained to lift the right hindleg (dCR) to stimulus dCS (rhythmic sound of a whistle) to avoid an electric shock to skin. Each dog of either alimentary or defensive group was trained separately. The training was conducted daily with 8–10 trials per session.

After 4–5 weeks of training, each intermittent conditioned stimulus used in the training was tested in a different situation. In order to extinguish the orienting reaction (such as sniffing and examining the environment) in the new situation, the test was always preceded by several habituation sessions in the test situation. There were four test situations.

The Neutral Test Situation

This test situation was a large empty room with no furniture except for a small table and a chair for the experimenter. The dog was allowed to move around freely. Neither conditioned nor unconditioned stimuli had been used here before.

In the alimentary group of 5 dogs, first a control neutral stimulus (nS), the sound of a rattle (never used before with these dogs) was applied for 15 seconds. This stimulus evoked only a slight orienting reaction. A few minutes later, the alimentary test stimulus (taCS), the sound of a flute (to which the alimentary instrumental CR was established in the uE before) was applied for 15 seconds. In all dogs, the taCS evoked only a motor reaction such as approaching the flute and sniffing it, or coming to the experimenter and placing the forepaw on her knees. However, *the instrumental reaction previously trained to this stimulus in the uE was not performed.*

In the defensive group of dogs, first a control nS, the sound of a rattle (never used with these dogs before), was applied for 15 seconds, evoking only a slight orienting reaction. Then the test CS (sound of whistle in two dogs, and flashing light in two other dogs) to which the defensive CR was established in the usual situation before, was applied for another 15 seconds. It was found that the test stimulus produced general anxious movements including shaking head (in case when electric shock to the ear was used as reinforcement in the uE). Nevertheless, the previously trained avoidance reaction (dCR) was not performed.

In both groups of dogs, therefore, neither alimentary nor defensive CS tested in the new neutral situation were able to evoke the previously trained instrumental CR, but still elicited a general motor reaction (classical conditioned reflex) related to the type of reinforcement used previously.

The Homogeneous Test Situation I

In this series, the test situation was the same large empty room that was used in the neutral test situation. However, the reinforcing stimuli (food for the alimentary group, and electric shock for the defensive group), were given in this situation previously.

In the alimentary group, the test was carried out on the five dogs previously used in the neutral test situation. In the first two sessions each dog was allowed to move freely and received a piece of meat every 20–30 seconds. Food was either thrown on the floor (for 2 dogs) or was offered in a bowl placed on a wooden beam lying along the wall of the room (for the other 3 dogs). Then during the third session, a new control stimulus, a certain type of noise, was applied for 15 seconds, producing no distinct reaction. One minute later, a conditioned stimulus, the sound of a flute (previously used in the training in the usual situation), was applied for 15 seconds. It turned out that the instrumental reaction previously trained in the uE *was not performed* by either of the two dogs which obtained food on the floor; one of them came to the experimenter and performed a "begging" movement. On the other hand, the three other dogs that obtained food in the bowl placed on a wooden beam *did perform the aCR* (placing their forepaw on the beam), previously trained in the uE.

In four dogs of the defensive group, first the control stimulus (the sound of a rattle) was applied for 15 seconds producing no distinct reaction. Then, one and a half minutes later, the test conditioned stimulus (flashing light for two dogs, and the sound of a whistle for two other dogs) was applied for 15 seconds. The avoidance reaction, previously trained in the uE, was not performed by any of the dogs, but shaking the head and anxious movements were observed. Then, an electric shock was given and the procedure was repeated. As before, there was not a clear reaction to the control stimulus (the rattle), but the application of the test stimulus elicited touching the ear (to which the shock was previously given in the uE) in one dog. only when, in addition, a band had been fastened on his leg, did this dog perform the previously trained movement of lifting this leg to the test stimulus. No previously trained reaction was performed by another dog, even when the band was fastened to its leg. However, two other dogs did perform the previously trained reaction when the test was preceded by an electric shock.

These experiments showed that the previously used conditioned stimulus elicited only a general defensive reaction but not the instrumental reaction previously trained in theusual situation. Only when some facilitating elements were present, such as a beam (for the alimentary group) or an electric shock (for the defensive group) would the previously trained CR appear.

The Homogeneous Test Situation II

In two dogs the alimentary reaction of lifting the right hindleg to the beating of the metronome was trained in the usual situation (uE). On the same days, another alimentary reaction of lifting the left foreleg in one of these dogs, and begging movement in the other dog, to the tactile stimulus, were trained in the test situation. When the metronome beating was applied in the test situation, none of these dogs performed the reaction previously trained in the usual situation, i.e., lifting the right hindleg. Instead, both dogs performed the movement trained in the test situation (i.e., lifting the right foreleg in one dog and begging movement in the other dog).

Two dogs of the defensive group were trained to lift their right hindleg to the flashing light in the usual situation, and to stand on hindlegs to the sound of a buzzer in the test situation; both movements secured the animal against the electric shock applied to the left ear. Then each conditioned stimulus was tested in the other situation. It was found that *neither stimulus evoked the instrumental reaction which had been trained to it before in the uE*. Instead, it elicited the reaction which had been trained to a different stimulus in the test situation, i.e., standing on hindlegs to the sound of a buzzer in one dog, and lifting the hindleg to the flashing light in the other.

The experimental results obtained in the homogeneous situation II in both alimentary and defensive groups of dogs, showed that the conditioned stimulus, previously used in the training of a definite instrumental reaction, does not evoke this reaction when it is applied in a different test situation. Instead, it evokes the instrumental movement which had been trained in the test situation to a different conditioned stimulus. In other words, the conditioned stimulus tested in a different situation evokes the reaction related to this new experimental situation but *not to the conditioned stimulus used in the previous training*.

The Heterogeneous Test Situation

In three dogs, the basic training was conducted in two different situations, one with alimentary reinforcement (situation A) and the other with defensive reinforcement (situation D). Situation A was a usual Pavlovian chamber. Situation D was arranged on the floor in an empty part of a room located in another building; the experimental part of situation D was separated from the observation section with a wooden partition. In both situations, the recording bands were fastened on all four legs of the dogs.

In situation A, the sound of a metronome for two dogs and a moving black disc for the third dog were used as alimentary conditioned stimuli; the instrumental reaction was placing the right foreleg on the feeder platform (aCR). In situation D, the sound of a whistle was used as a defensive CS; the instrumental reaction was lifting the right hindleg to avoid a shock to the same leg (dCR). The training was conducted in alternate series each lasting for several days in each situation. Then the test was performed in which the defensive CS was applied in the alimentary situation, and the

TABLE 5.1
Results of Training of Two Various Instrumental Conditioned Reflexes,
CS1-CR1 and CS2-CR2, in Various Situations

Kind of situation	Trained reactions	Results
One situation	CR1, CR2	Both CR1 and CR2 often appear to each CS or one replaces the other
Two similar situations	CR1, CR2	Same as above
Two different situations	CR1, CR2	Each reflex appears only to the CS to which it was trained in the given situation

TABLE 5.2
Testing an Instrumental Conditioned Reflex, CS1-CR1 in Various Situations

The kind of the test situation	Results		
	General reaction	Correct reaction	Other reaction
Neutral	yes		
Homogeneous in which UCS was previously used	yes	yes (in the presence of facilitating elements)	
Homogeneous in which other reflex, CS2–CR2 was trained	yes		yes
Heterogeneous	yes	yes (in the presence of facilitating elements)	

alimentary CS was applied in the defensive situation. The results of this procedure are described below.

The Application of the Defensive Conditioned Stimulus in the Alimentary Situation

When the defensive CS was applied in the alimentary situation for the first time, one dog performed a slight but distinct lifting of the right foreleg (defensive CR). Another dog reacted to the defensive CS with a slight lifting of the right foreleg (partial aCR), followed by shivering, bristling of hair, and anxious behavior. No clear reaction to defensive CS was seen in the third dog. When the defensive CS was repeated in the next trial, all three dogs performed the lifting of the right hindleg (dCR). When a control stimulus, the sound of bubbling water (never used with these dogs before) was applied, a slight orienting reaction only was observed in one dog, while two other dogs performed an alimentary CR. When the control stimulus was repeated, all three dogs performed the alimentary CR.

The Application of the Alimentary Conditioned Stimulus in the Defensive Situation

The first application of the alimentary CS in the defensive situation did not produce any motor reaction in two dogs, but the third dog performed 3 slight movements of the right hindleg (i.e., the defensive movements). The repetition of the alimentary stimulus in this situation again did not produce any effect in the first two dogs. The third dog, however, first lifted its right hindleg and then its right foreleg and held it in this position for several seconds; this resembled the aCR trained in the alimentary situation.

Summarizing the results obtained in heterogeneous situations it can be said that an alimentary CS applied in the defensive situation, and a defensive CS applied in an alimentary situation evoked only a general conditioned reaction connected with each stimulus or did not evoke any reaction at all; only in a few cases an adequate instrumental reaction (or one similar to it) was observed. (See Table 5.1 and Table 5.2).

6

Examples of Complex Conditioned Reflexes to the Situation

Although the famous experiments of Pavlov's laboratory were based strictly on a single external stimulus and a single conditioned reaction, other investigators used complex stimuli and complex reactions in their studies on conditioning and learning.

The most eminent cases of complex conditioning were provided by Thorndike (1911). As already mentioned in chapter 2, Thorndike conducted his experiments on various animals such as fish, chicks, cats, dogs, and monkeys. His studies provided many examples of complex stimuli as well as complex motor behaviors. In spite of the fact that Thorndike called the topic of his experiments "trial-and-error learning," the behaviors he studied can be regarded as cases of the complex instrumental conditioning.

Detour Behavior

More recent studies of complex instrumental conditioning include a topic such as detour behavior. A "detour" or a roundabout behavior," the ability to go around an obstacle on the way to a reward, was considered a result of "insight" by Kohler, author of the famous book *The Mentality of Apes* (Kohler 1925). A similar view was expressed by other investigators who conducted their observations on rats (Higginnson 1926, Hsiao 1929, Tolman and Honzik 1930); birds (Teyrovski 1930, Lorenz 1932, 1939); reptiles (Fisher 1933); and fish (von Schiller 1942). However, Thorpe (1956) emphasized that before accepting an explanation of detour behavior as a result of insight, one must first determine whether this behavior appears in an untrained animal as its first response.

The problem of detour behavior was also studied by Beritashvili and his associates (Beritashvili 1941, Beritashvili and Akhmeteli 1941) who conducted their observations on dogs, rabbits, hens, and pigeons. The results of their studies led them to a conclusion that the animal performs the detour behavior due to a psycho-neural process which reproduces a picture of the external world in the animal's mind thus guiding its movements towards the food. Later, Roginski and Tikh (1956) studied detour behavior on chickens, rats, and monkeys, and found that monkeys learned to make the detour fastest, and chickens most slowly; they concluded, therefore, that the detour reaction depends on the phylogenetic level of the animal. Pavlov (1949), however, expressed the opinion that the detour behavior has an acquired, conditioned reflex characteristic.

In order to gather more data concerning the mechanisms involved in the performance of detour behavior, systematic studies were conducted by Wyrwicka (1959).

In the preliminary testing period of this study it was found that adult dogs easily made a detour when food was placed behind a visible obstacle. A question arose, therefore, as to whether this reaction was inborn or acquired by experience. In order to answer this question, a series of experiments was performed on puppies that had been kept in a controlled environment from birth and never had any occasion to make a detour before.

The study was conducted on 28 puppies, male and female, 12–131 day old, of 8 different litters from 5 female and 8 male dogs. Mother and puppies of each litter were housed in a large pen, 2 x 3.5 m at the bottom and about 1.5 m in height. Their home compartment was empty and the pups had no occasion to go around any object. They were first nursed by the mother, and later they were fed solid food (mixture of cooked meat and cereal) with the mother.

The tests were carried out separately for each pup, in a large, empty room (different from the pen). The experimental space was partially divided by a wire-net partition as schematically shown in fig. 6.1.

FIGURE 6.1

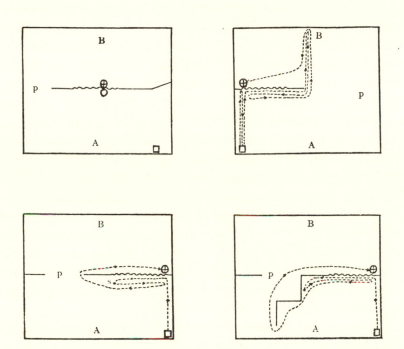

Samples of detour behavior of puppies (1–4 months old) that were controlled from birth and never had occasion to make the detour before. (Photocopies of sketches of detour behavior of single puppies, made by the experimenter during the test session).

Straight line, wood partition; wavy line, wire-net partition; square at the bottom of each sketch, starting place; empty circle at the front of wire-net, site of the bowl and milk during preparatory sessions; circle with a cross, site of the bowl behind the wire-net partition during the test session; the broken line with small arrows traces of detour behavior.

In the preliminary series of 5 sessions (each lasting 15 minutes), each of 12 puppies of the first group was allowed to walk in both parts of the experimental compartment (i.e., before and behind of the partition); 11 other pups of the second group were allowed to remain only in front of the partition. The third group of 5 pups was trained in a space arranged in the garden with a wire-net partition (similar to that used in the room) where they were allowed to run both in front of and behind the wire-net.

Then the testing series of experimental sessions was performed, separately for each pup. At the beginning of each session, the pup was offered food in a bowl placed in front of the wire-net partition. After approximately 20 daily sessions, the bowl with food was placed *behind* the partition, just across from the previous location of the bowl. The food was clearly visible to the pup, but it could not be reached from the front of the partition. In order to get the food, the pup had to go first to the end of the partition (i.e., to go away from the site where the food was placed behind the fence) and then make a U-turn toward the food (see sketches in fig. 6.1).

It turned out that most of the 12 pups of the group that had been allowed to walk in both parts of the experimental compartment before the test easily found the way to the food in seconds; only a few pups in this group needed more time to make a detour. Similar results were obtained on 5 pups that were allowed to run both in front of and behind the partition in their preliminary training in the garden.

On the other hand, *none* of the 11 pups that had been allowed to remain only in front of the partition before the test could find the way to the food. Independent of age, they only repetitively scratched the partition at the site where the food was visible behind the wire-netting. This behavior was observed for several consecutive sessions until finally, earlier or later, each pup started to walk back and forth alongside the net, and accidentally found the passage to the bowl (see a schematic pathway of a pup in fig. 6.1). On an average, 4–6 sessions were needed for each pup to learn the detour. It is interesting that age did not influence the response. When the test was conducted on 4-month-old pups, their learning of the detour took even longer than learning in the 2-month-old pups.

Once the pup learned to make a detour in a room, it was able to make it also in a different surrounding, namely in the garden with the use of the same (or similar) partition and the same procedure. This means that the situational stimuli were recognized by the pup as the same as those used in the room before, and thus, the detour reaction was performed.

In order to gather more information concerning the nature of this reaction, an extinction test was performed separately on each of 3 pups. After the establishment of the detour behavior, the experimental sessions were continued but *the food was withheld*, i.e., the bowl behind the partition was *empty*. There were 3 trials per session. The pups first continued to run to the bowl behind the partition, but after 10 sessions in 2 pups and 30 sessions in the third pup, they stopped going around to the bowl. After the extinction of the detour reaction in the room, this behavior was also absent in the garden.

It was concluded, therefore, that the detour behavior *is not a result of "insight."* Neither is it an inborn unconditioned reaction. Instead it is a behavior, independent of age, acquired through individual experience, i.e., a *complex instrumental conditioned reflex.*

In this case, both the stimulus amid the reaction are complex. The stimulus is composed of permanent situational elements such as visual stimuli of the wire-net partition, the bowl, and the passage. The motor reaction involves both excitatory and inhibitory processes. First, the initial motor reaction to the bowl (scratching the partition in front of the animal in an attempt to reach the food behind the partition) had to be fully extinguished, i.e., inhibited by not obtaining the reinforcement (food). Then only the correct motor reaction, a U-turn, rewarded by food, can gradually develop (fig. 6.1).

Of course, there are more instances of behavior where both stimuli and reactions are even more complex. The manipulatory reactions of animals of various species described by Thorndike (1911) and chain reflexes analyzed by Skinner (Skinner 1938, Ferster and Skinner 1957) may serve as examples of complex conditioned behavior. The directional instrumental reactions studied by Lawicka (1959) and Konorski and Lawicka (1959) also belong to this category.

Food Reinforcement as a Conditioned Stimulus

Let us try to analyze the meaning of food as an important reinforcing stimulus in the conditioning process. Usually, little attention is paid to food which is so commonly used in a variety of conditioning techniques. The attention of the experimenter is usually concentrated on the animal's activity preceding the food reinforcement.

However, some studies in our laboratory showed that the act of eating can be considered an *instrumental conditioned reflex*. In experiments of Wyrwicka and Chase (1972), cats with electrodes implanted in the lateral hypothalamus were offered a milk and broth mixture in two identical containers. Drinking from one container was rewarded by desirable electrical stimulation in the hypothalamus (the desirability of obtaining the hypothalamic stimulation was first tested by the method of self-stimulation arranged so that the cat pressed a lever to get that stimulation). Drinking from another container was not reinforced by pleasurable hypothalamic stimulation. It appeared that cats concentrated on drinking milk-broth mixture only from the container with stimulation and ignored the same mixture in the other container. In other experiments, cats learned to eat slices of bananas or tasteless and odorless jellied agar when this was rewarded by hypothalamic stimulation (Wyrwicka 1978, 1981, 1988).

Feeding behavior, therefore, should be considered a *complex conditioned act*. It is usually preceded by a number of classical and instrumental conditioned reactions which prepare the animal for obtaining food. The classical type reactions include secretion not only from salivary glands but also from various internal organs (see Bykov 1957) as well as changes in respiration and circulation (Gantt 1960, Gantt and Hoffmann 1940) and some unconditioned motor activities such as orienting and investigatory reactions. Instrumental preparatory behavior depends both on the previous experience with food and on the present circumstances. In the laboratory, the preparatory instrumental reaction is simply a trained movement, easy to observe and record. In nature, the preparatory activity may be more complex, such as seeking food, hunting or fighting for food. In various circumstances, this behavior may start a long time before eating. For instance, in human life, the preparatory feeding behavior may include all activities related to obtaining food such as shopping for food, preparation of a meal, etc.

FIGURE 6.2

PREFEEDING ————————→ **ACT OF EATING**

Instrumen. CRs

(preparation of meal or
going to restaurant, etc.)

Complex or Oral and
Other Instrumen. CRs

(masticatory and other CRs
related to the kind of food)

A self-explanatory diagram of activities occurring before and during the
act of eating.

Let us now analyze the instrumental type behavior related to
alimentary behavior. This behavior is composed of two parts as
shown in a diagram in fig. 6.2. the first part of this diagram shows
the preparatory instrumental type activity preceding eating such
as performance of the previously trained movement (in the laboratory) or a complex of motor reactions such as seeking food or
hunting, etc. (in nature).

The second part of the feeding behavior is the act of eating itself. A closer look at this act shows that it is quite complex and
dependent on the kind of food to be consumed. In mammals, the
first feeding behavior is sucking. The fact that it appears early
after birth indicates that sucking behavior is inborn (although it is
not excluded that it can occur even before birth in the womb by
sucking the surrounding fluids). Later in life, the young start to
follow the mother or other adults of their colony in seeking food
and eating that food (Galef 1977, 1978; Galef and Clark 1972).
This suggests that the motor ability to eat various kinds of food
(other than mother's milk) is acquired by experience later in life.
This ability is based on a number of inborn reflexes, such as mas-
tication, i.e., the movements of opening (digastric reflex) and closing (masseteric reflex) the mouth (Clemente et al. 1966; Chase
and McGinty 1970 a,b; Chase and Babb 1973) as well as the movements of the tongue and other parts of the mouth cavity. The motor performance may vary depending on the kind of food to be
eaten. The motor (instrumental) reaction of eating a piece of meat
is different than the motor reaction of eating creamy pudding or
drinking milk. Due to the repetitive experience with various kinds
of food, the act of eating gradually becomes more and more com-

plex. Even later in life some further adjustments in the act of eating may take place when the unknown kinds of food are being offered. Therefore, the food, which is commonly called an "unconditioned stimulus," turns out to be a specific *conditioned stimulus*, and the act of eating becomes a *conditioned instrumental reaction*. In other words, the alimentary instrumental conditioned reaction is reinforced not by food as an unconditioned stimulus but by *food as a conditioned stimulus* when it is seen or sniffed by the subject. In this case, only the final effect of eating, the sensory satisfaction resulting from food consumption, can be considered the "reward" or the "reinforcement" (Wyrwicka 1975).

A question can be asked whether or not the act of eating can depend on the visual or olfactory stimuli related to a particular portion of food, independently of the situation where this food has been offered. It seems that environmental factors may still influence the reaction of eating. One of these factors is the state of satiation. In this case, the act of eating may not occur. In another case, the unsatisfactory environment such as a dirty tablecloth or the presence of a constant loud noise, etc., may inhibit the consumption. There are many such factors restraining eating.

On the other hand, there are also factors that facilitate eating. An example is the already quoted study by Wyrwicka and Chase (1972) in which cats were offered food in two containers; eating from one of these containers was rewarded by desirable hypothalamic stimulation, while eating from the other container was not. The cats chose to eat only from the container where they obtained the pleasant stimulation.

Experiments with completely satiated goats which refused to eat any more can serve as another example. These animals, however, started to eat again when they obtained a pleasant light electrical stimulation in the "feeding center" in the lateral hypothalamus each time they ate, even when this was leading to overeating (Wyrwicka et al. 1959, 1960, 1982, 1988).

Presleep Behavior

A case of conditioning to the situation was observed in cats in relation to sleep. As it was demonstrated by Sterman and Clemente (1962), Clemente et al. (1963), Yamaguchi et al. (1963), Clemente (1968), McGinty and Sterman 1968, and others, electrical stimulation of the basal forebrain area (BFA) produced sleep accompanied by specific changes in the electroencephalic activity (EEG) of the brain. In a more recent study by Wyrwicka and Chase (1994), experiments were conducted on three chronic, unanesthetized, undrugged cats, bearing monopolar electrodes implanted in the basal forebrain area. Light electrical stimulation (2–4 V, 50–100 Hz, 1 ms dur/imp, 0.5 s trains at 1/s frequency, for 2–5 minutes each time, followed by a nonstimulation period of the same duration) resulted in producing synchronized EEG spindles which were accompanied by the cat assuming the posture typical for presleep behavior such as crouching with the head down, and the eyes partly or fully closed. After several sessions with the use of BFA stimulation, two of the cats started to demonstrate presleep behavior almost immediately after entering the experimental compartment, ignoring the food (which was always present in the experimental compartment). The third cat usually ate some food and afterwards began to show presleep behavior. When stimulation was withheld in the extinction procedure, the cats still continued to exhibit the presleep behavior in the absence of the stimulation during several consecutive sessions. Thereafter their usual behavior (of the pre-stimulation period) gradually returned, including consumption of the offered food. The authors concluded that repeated BFA stimulation produced conditioning of the presleep behavior to the complex of environmental stimuli of the situation.

7

Conditioning of the Activity of the Internal Organs

Intensive studies were undertaken by N.E. Miller and his associates on the possibility of producing instrumental conditioning related to internal organs innervated by the autonomic nervous system. Miller and DiCara (1967) demonstrated that rats immobilized by an injection of curara and artificially respirated learned either to increase or decrease their heart rate (dependent of the experimental design) to obtain the rewarding electrical stimulation in the basal forebrain bundle (cf. Olds and Milner 1954; Olds 1962). In other experiments, the curarized rats learned to increase their heart rate (or decrease it, dependent on the chosen experimental procedure) to avoid an electric shock to the skin (DiCara and Miller 1968a). Similar results were reported by Throwhill (1967).

Using the same method, it was possible to change the vasomotor activity, i.e., to obtain either vasoconstriction or vasodilation, when the required activity was reinforced by desired electrical stimulation in the basal forebrain bundle (DiCara and Miller 1968b). Another study showed that thirsty, noncurarized dogs learned to increase salivation when rewarded with access to water; it was observed that salivation in these dogs increased with the training (Miller and Carmona 1967). It was also possible to increase the rate of urine formation in one group and decrease it in another group of curarized rats when electrical stimulation in the basal forebrain bundle was offered as a reward (Miller and DiCara (1968).

A question arose as to whether the obtained instrumental autonomic responses can be entirely independent of the central nervous system. In order to answer this question, a high dose of curara (4.8. mg/kg over 3 hours) was used, resulting in full stoppage of muscular activity; in spite of complete immobilization of the animal, rats still learned to change their heart rate (DiCara and Miller 1968c).

However, different results were obtained in experiments on dogs by Black (1967). This investigator studied the instrumental responses of the heart rewarded by withholding of shock, under various degrees of curarization. When a higher degree of curarization was used (4.1 mg/kg over 8 hours), resulting in a complete suppression of muscular activity, the possibility of obtaining instrumental conditioning of the heart rate strongly diminished. These results led to a conclusion that instrumental conditioning of heart rate may be mediated by some central processes which are in charge of motor activities (Black 1967).

Instrumental Conditioning of the Electrical Activity of the Brain

Several studies have shown that electrical activity of the brain (EEG) can be conditioned, i.e., it becomes an instrumental conditioned reaction. Attempts of conditioning of the electroencephalic function of the brain were made by various investigators. One such study was that of Olds and Olds (1961), who recorded unitary discharges from paleocortical structures of the rat's brain. Each time the number of discharges from the unit increased, the rewarding electrical stimulation of the medial forebrain bundle was applied. As a result, the number of discharges became considerably greater.

In another group of experiments conducted on cats, Izquierdo et al. (1965) studied the possibility of establishing a conditioned EEG rhythm to a tone applied while the animal was sleeping. The tone was repeatedly applied until EEG arousal was no longer elicited by the tone, i.e., until full habituation had been achieved. Next, the same tone was applied for 4 seconds; two seconds after its discontinuation, a weak electric shock to the cat's skin was applied. After a number of such trials the tone alone began to evoke

EEG arousal. Each time the EEG reaction occurred to the tone, the shock was withheld. After a number of repetitions of this procedure, EEG arousal began to appear in each trial while the cat remained lying down and apparently asleep as shown by electromyographic recording (EMG) from the animal's neck muscles. In this experiment desynchronization of the EEG pattern became an instrumental avoidance reaction.

The problem of conditioning of EEG waves was also studied by Carmona in the laboratory of Miller (Miller 1969). In one group of cats, the appearance of high amplitude slow EEG was each time rewarded with the electrical stimulation of the medial forebrain bundle, whereas in another group a low amplitude EEG activity was rewarded in the same way. As a result, the cats of each group learned to produce more of the particular rewarded rhythm.

Similar studies were also conducted on human subjects by Kamiya (1962, 1968). In his experiments, appearance of the alpha rhythm from occipital cortex leads was each time reinforced by a small monetary reward. This procedure resulted in a significant increase of the alpha rhythm. Other studies include those of Nowliss and Kamiya (1970) as well as experiments of Rosenfeld et al. 1969) and Brown (1970).

An attempt was also undertaken to condition, in cats, the alpha-rhythm (high- amplitude waves, 8–12 cycles per second) appearing in parieto- occipital cortical leads. Each time such a burst appeared, a small food reward (piece of meat) was given to the cat. After a number of such trials alpha activity began to appear each time the tone was applied (Wyrwicka 1964).

These first observations on conditioning of EEG activity should be considered as preliminary results of studies undertaken later with the use of precise techniques which included both better experimental control and detailed analysis of the results. Namely, in a study conducted on cats, an EEG activity called "sensorimotor rhythm" (SMR) of high amplitude waves of 12–14 cycles per second from the coronal gyrus of the cortex (Sterman and Wyrwicka 1967, Wyrwicka and Sterman 1968, Sterman et al. 1969) was used. Each time a burst of SMR appeared an automatic switch delivered a small portion of milk to the cat. After a number of trials, the

SMR burst appeared more often and at regular intervals (Wyrwicka and Sterman 1968). When a milk reward was withheld during the extinction procedure, the frequency of the SMR bursts temporarily increased at the beginning of extinction. The finding of a temporary increase of the instrumental reaction, which is usually observed at the beginning of the extinction process (see Konorski 1948,p. 217; 1967, p. 361), supports the view that the sensorimotor rhythm became conditioned as an instrumental type reaction.

Further research focused on precise localization of the brain region where the sensorimotor rhythm is initiated. Research of Sterman and Clemente (1962), Clemente and Sterman (1963, 1967), Clemente et al. (1963), Sterman and Wyrwicka (1967), Sterman et al. 1969) revealed that this region is situated in the basal forebrain area (BFA). According to the stereotaxic atlas of the cat's brain of Jasper and Ajmone- Marsan (1960), the coordinates for this area are A 13.0–15.0, L 2.5–3.0, H—4.0, although some differences from cat to cat may exist. Bilateral lesions in the basal forebrain area in cats resulted in suppression of sleep (McGinty and Sterman 1968).

Electrical stimulation of the basal forebrain area leads to the production of presleep behavior such as cessation of eating and assuming a sitting or lying position with the head lowered and eyes half or fully closed. Usually one or two minutes later, spindle bursts of sensorimotor activity (SMR) appear on the EEG recording. Both the behavioral and EEG responses resemble those observed during natural presleep behavior.

Repeated stimulation of the BFA leads to conditioning of the SMR as shown in studies of Wyrwicka et al. (1962), and Clemente et al. (1963). In experiments on cats, the BFA stimulation period was each time preceded by a tone (T1). After several repetitions of this procedure, T1 alone started to evoke the spindle bursts of SMR. A control stimulation in the thalamus preceded by another tone (T2) in the same animal, did not evoke SMR; instead, a high frequency EEG pattern was observed. Repetitive stimulation of this thalamic area led to an appearance of high frequency EEG activity to T2, without any appearance of SMR. These experiments further supported the observation that electrically elicited EEG bursts typical of presleep stage can be conditioned.

Research conducted on human subjects suffering from insomnia showed that presleep behavior can also be conditioned. In a study of Poser et al. (1965), beats of a metronome were paired with a sleep- inducing barbiturate (methohexitone) once a day, for 16 days. Toward the end of this period, the subjects reported relaxation and soporific tendencies during metronome beats.

In a treatment used by Evans and Bond (1969), administration of a sleep- inducing barbiturate (methohexital sodium) was paired with self- produced counting to 28 as a conditioned stimulus. After eight weekly sessions with this procedure, an increase in the amount of sleep, from an initial 2 hours to a final 6 hours a night, was observed.

In a more recent study on human subjects by Caruso et al. (1988), classical conditioning of sleep onset was produced using a sleep- promoting hypnotic drug, triazolam, that served as an unconditioned stimulus. It was paired with pill ingestion, and a tone, together with the environmental stimuli, served as the conditioned stimuli. When, under the same conditions, lactose (a placebo) was substituted for triazolam, sleep latency remained shorter than the previously determined baseline.

Earlier studies on animals revealed the presence of a EEG slow-wave pattern from the sensorimotor cortex during behavioral inactivity or trained inhibition (Gastaut 1958, Kogan 1960, Anokhin 1961, John et al. 1961, Rowland 1961, Donhoffer and Lissak 1962, Roth et al. 1967, Sterman and Wyrwicka 1967, Wyrwicka and Sterman 1968). A systematic study on somatomotor suppression and excitation by stimulation of the orbital gyrus in unanesthetized freely moving cats was conducted by Chase and McGinty (1970 a, b). They found that orbital cortical stimulation resulted in the suppression of the masseteric reflex and the digastric muscle contraction; however, a gradual reduction of the effectiveness of the cortical stimulation was observed as the animal passed from the alert state into quiet sleep.

In experiments by Babb and Chase (1974), cats were trained to produce the sensorimotor rhythm (SMR) while the masseteric and digastric reflexes were monitored throughout the session. It was observed that during SMR small amplitude masseteric reflexes were depressed, whereas large amplitude masseteric reflexes were not. According to the authors' interpretation, the suppression of the masseteric reflex is related to a decrease of afferent excitatory input produced by reduced cortical discharge; on the other hand, strong reflex evocation would compensate for the reduced input, masking the inhibitory effect.

In a special study of Chase and Harper (1971), cats were trained to produce a synchronized EEG pattern of 12–14 cps of sensorimotor activity to get a milk reward. During the presence of SMR, somatic motor activity was inhibited; at the same time, heart rate decreased, and the respiratory pattern became more regular than during control periods. In a later study, neurons in the thalamus were found to discharge in a "burst- pause" manner during SMR, similar to that found in quiet sleep (Harper 1973, Harper and Sterman 1972). These data suggest that the sensorimotor rhythm is related to *motor inhibition*.

In all experiments described in this chapter a process of instrumental conditioning took place. In each case the subject was required to produce a definite reaction such as an increase or a decrease in heart rate, obtain vasoconstriction or vasodilation, increase or decrease the amount of salivation or urine formation, etc. In some other experiments an increase of neuronal discharges or an appearance of the EEG waves of a definite frequency was a conditioned reaction. Only after the performance of the required task was the reward given. In each case, the process of instrumental conditioning was performed on the internal organs of the body. In other words, this process took place in the *internal environment of the body*. This suggests that there exists not only an external situation in which the conditioning usually occurs, but also an *internal situation* which plays an important role in the life of the organism. This problem will be further discussed in the following chapter.

8

Therapeutic Role of the
EEG Feedback in Epilepsy

The elicitation of specific changes in functions of various internal organs such as heart rate, increase or decrease of salivation, and others described in the previous chapter, suggest that these changes have been acquired by conditioning. For instance, according to observations of Miller and Carmona (1967) thirsty dogs increased salivation with the training. Similarly, heart rate increased when the animal was rewarded by desired electrical stimulation in the basal forebrain bundle (DiCara and Miller 1968). These and other examples of changes in functions of the internal organs suggest that these events take place accordingly with the basic mechanisms of instrumental conditioning.

A question can be asked as to what instrumental method has to be used by the organism to achieve the required functional change. For instance, an increase in respiration can result in an increase of heart rate; similarly, the movements of the tongue or the jaws may increase salivation. In these cases the required instrumental reaction has been performed with the use of another reaction (respiration, movements of jaws, etc.).

However, these instrumental reactions occurred also in immobilized (curarized) animals. In such cases, the motor activity of one system could not be used to produce the reaction in another system. In spite of this, the animal somehow finds the means leading to the performance of the required behavior.

Especially intriguing is the achievement of a required pattern of electrical activity of the brain. Some such cases as described in the previous chapter demonstrate that with the use of a specific reward, it is possible to obtain the required EEG pattern both in animal and human subjects (Olds and Olds 1961, Miller 1969 / experiments of Carmona/, Kamiya 1962, Wyrwicka 1964, and others).

The development of studies on conditioning of electrical activity of the brain led to the question as to whether it would be possible to use some of the evoked EEG patterns in a therapy of certain neural disorders. One such disorder is epilepsy.

Treatment of Epilepsy with the Use of Sensorimotor Training

Epilepsy can be described as a state of disregulation of the function of the motor system, characterized by convulsions (seizures), often with a temporary loss of consciousness. The EEG recording in epileptics showed a completely disordered picture of the electrical activity, which was replaced by slow regular waves. These slow waves were seen in epileptics only in the absence of seizures and were observed mostly in the sensorimotor cortical area. These waves were described in the previous chapter as a "sensorimotor rhythm" being related to motor inhibition.

The finding of the inhibitory character of the sensorimotor rhythm led to the question of whether it would be possible to use the SMR as a natural treatment in conditions involving disregulation of the motor system in epilepsy in humans.

In order to answer this question, the following study was conducted on an epileptic by Sterman and Friar (1972). The subject was a 21-year old female with a history of convulsive disorder which began when she was 16 years old. The onset was insidious and consisted of nocturnal, generalized major seizures. There was no family history of epilepsy. Neurologic examinations failed to demonstrate any localized lesions. The EEG recordings taken three years before and just prior to the study showed a generalized spike-wave activity in the left fronto-parietal area of the brain. The seizures were preceded by a non-specific aura, but there were no localizing features of any precipitating factors. Daytime incidents were limited only to some small movements such as wrinkling of the brow associated with the left lateral deviation of the eyes, crossing of the right arm to the left knee, and falling to the left. The majority of seizure incidents were currently nocturnal, occurring in the later portion of a night's sleep.

The frequency of incidents varied from 2 to 4 per month at the onset to 1 per 3 months one year later. During the 12-month period immediately preceding the treatment, seizures were present irregularly at an average rate of about 2 per month. The subject had variable success in using a combination of drugs such as Dilantin, Mysoline, Peganone, Dianox, and Ritalin. In the preceding 12 months she also daily used Dilantin and Mebarol.

The training techniques included the EEG recording from the scalp sensorimotor and occipital areas, using needle electrodes. Placement of electrodes was standardized by reference to the international (10–20) system. The electrocardiogram (EKG) and electromyogram (EMG) were also recorded. During the sessions the patient was sitting comfortably on a reclining chair. The instructions were delivered through standard tape recording. The device recording the SMR activity (feedback) consisted of a rectangular unit suspended from the ceiling several feet above and in front of the patient. The unit contained two rows of ten small lamps covered with transparent colored Plexiglas. When the EEG activity of appropriate amplitude and duration appeared, the lamps in the top row were successively lighted; each advance was accompanied by the sounding of a single chime which was used as a reward. With the eleventh successful EEG response the lights of the top row were extinguished and the lamps of the bottom row were activated; each time this was accompanied by a double chime reward. Usually, 200 to 300 rewards were obtained during a single session lasting from 30 minutes to one hour. The sessions were conducted once a week during the first month and then twice during the following three months.

The results of this treatment were dramatic. Initially, the patient showed a moderately low level of SMR feedback. Later, however, after adjustment of the reward from 12–14 cycles/sec (c/sec) to 11–13 c/sec, a gradual increase in the production of SMR feedback was observed. After three sessions there were no seizures for the following period of 3 months, except one mild nocturnal seizure. At the same time, some changes occurred in the personality of this patient. Being previously a quiet and rather timid, non-assertive individual, the patient progressively became more outgoing, with personal, confidence and even an enhanced interest in her appearance. She also reported a shorted latency to sleep onset, and a more rapid awakening in the morning. Moreover, this patient voluntarily ceased her previous practice of supplementing anticonvulsant medication while experiencing an aura, without seizure, and reported a reduction in such occurrences. In a later phase of the study she obtained an additional small monetary reward for producing SMR; this resulted in the highest rate of performance she had ever showed.

A reduction of epileptic seizures following the training of 12–15 c/sec SMR feedback was confirmed in further studies of Sterman and his colleagues (Sterman et al. 1974, Sterman and MacDonald 1978, Sterman and Shouse 1980) as well as in the studies of Finley et al. (1975), and Seifert and Lubar (1975). However, several other authors who had confirmed seizure reduction challenged the concept of a specifically therapeutic 12–15 c/sec EEG rhythm. Some investigators reported that the enhancement of this EEG pattern was not always accompanied by a decrease in seizures (Kaplan 1975, Wyler et al. 1976, Kuhlman and Allison 1977, Kuhlman 1978, Cott et al. 1979).

In order to obtain more data, the following study was undertaken on 8 poorly controlled seizure patients by Sterman and MacDonald (1978). Four of these patients were rewarded alternately for facilitation and suppression of 12–15 c/sec SMR activity. All these patients showed a seizure reduction only during positive reward for the 12–15 c/sec pattern. Four other patients were rewarded for the facilitation and then suppression of 18–23 c/sec rhythm. Three patients of this group showed significant seizure reduction.

A further study by Sterman and Shouse (1980) was conducted on another group of 8 epileptic patients who were not adequately controlled by anticonvulsant medications. The patients ranged in age from 18 to 35 years, and their history included diagnosed epilepsy from 8 to 24 years. The techniques used for this group included not only the training with the use of various frequencies of waves but also sleep EEG spectra, clinical EEG and anticonvulsant blood levels. The training was conducted mainly in the patient's home and lasted 3 months.

The investigators found a compliance with training instructions and an acquisition of the responses. Overall anticonvulsant blood levels were low and unrelated to the EEG or seizure changes. The clinical EEG corresponded to sleep EEG and seizure rate outcomes. Power spectral analysis of samples of "non-rapid eye movement" (non-REM) sleep from all-night EEG recordings in the first subgroup of patients corresponded to the pattern of seizure rate changes in this group. EEG changes were also limited to sensorimotor cortex in the second subgroup, but were linear, and paralleled a progressive decrease in seizure rate. Both groups showed the same pattern of EEG changes with seizure reductions. Low and high frequency waves were reduced and intermediate frequencies were increased. The correlation analysis confirmed this relationship. The pattern of these changes suggested a normalization of sensorimotor EEG substrates related to the EEG feedback training.

The results of studies described above, confirmed by other investigators (Finley et al. 1975, Kaplan 1975, Wyler et al. 1976, Ellertsen and Klove 1976, Kuhlman 1978, Kuhlman and Kaplan 1979, Wyler et al. 1979). Cott et al. (1979), emphasized the dependence of the results on the procedure while Lubar et al. (1981) stressed that the severity of epilepsy symptoms may influence the EEG feedback therapy.

A study by Lantz and Sterman (1992) confirmed the observation of Lubar et al. that the effect of feedback training was indeed dependent on the severity of symptoms which could include a variety of cognitive, motor, and psychosocial functions. An improvement in these functions could occur only in subjects who were successful in learning the trained response and reducing seizures.

EEG feedback activity is a unique case of conditioning which occurs entirely inside of the internal situation. This kind of conditioning is produced by an appearance in the brain of a complex pattern of motor excitation produced by a specific physiological state which normally leads to epileptic seizures. The occurrence of this pattern of excitation is a conditioned stimulus which evokes a specific conditioned reaction. The conditioned reaction in this case consists in low-frequency, high amplitude EEG waves (i.e., the sensorimotor rhythm, SMR). The appearance of this rhythm indicates that the state of motor inhibition has taken place leading to the disappearance or prevention of the seizures.

The acquisition of this reflex which should be considered a defensive instrumental reaction fully depends on the subject. The patient must learn to recognize a change in sensations preceding the occurrence of epileptic seizures and immediately initiate an inhibitory action (by producing SMR) thus preventing the attack. If the patient acquires this reaction by practice, the seizures may be avoided each time. The clinical records described above show that such control of epilepsy is possible to achieve.

9

Theoretical Comments

According to experimental results, the relationship between the intermittent stimulus (CS) and the situation in which it is applied can be described as follows.

When two instrumental reactions to two different intermittent conditioned stimuli (one acoustic and the other visual), respectively, were trained in *one situation*, each CS often elicited the reaction related to the other CS, or both instrumental reactions. The same results were obtained when the training was conducted in *two similar situations*. However, when one of these reactions was trained in one situation and the other in another situation, *completely different* from the other one, only one instrumental reaction (CR) was obtained in each situation (that which was trained in this situation) (see the summary in Table 5.1).

When each intermittent CS was separately applied in various situations, the following results were obtained.

An application of the CS in the *neutral situation* (in which neither an unconditioned [US] nor a conditioned stimulus was even applied) did not evoke a specific CR, but only a general reaction which contained some elements of alimentary or defensive reactions connected with this stimulus in the previous situation.

An application of the CS in the *homogeneous situation* (in which no instrumental CR was ever trained but the same unconditioned stimulus, food or shock, was used in the previous situation), evoked general alimentary or defensive behaviors. In some cases a truncated or delayed instrumental reaction was performed, especially when some element of a previous situation, such as a beam resembling the platform, was present, or when an unconditioned stimulus (electric shock) was given.

An application of the CS in *another homogeneous situation* (in which an other instrumental reaction had been previously trained to a different CS) evoked this other instrumental reaction.

An application of the CS in the *heterogeneous situation* in which an alimentary CS was tested in a defensive situation (where a defensive reflex was previously trained), and vice versa, a defensive CS was tested in an alimentary situation (where an alimentary reflex was previously trained), evoked a general conditioned reaction, and only in a few cases (especially when the test stimulus was repeated), a correct instrumental reaction appeared to the tested stimulus.

These results are summarized in Table 5.2.

The experimental data show that a single conditioned stimulus produces the definite instrumental CR only in the situation in which it was trained. *Once the situation has been changed, the conditioned stimulus is not able to produce the correct instrumental conditioned reaction.* Sometimes, a diminished and modified reaction may be present, hardly resembling the correct CR.

On the other hand, the conditioned stimulus tested in various situations always elicited general activity with some features related to the unconditioned stimulus previously used in the training. In the case of a defensive CS, it was a general defensive reaction, e.g., crouching, shaking the head, touching the ear (with paw) where the electrodes were attached before, or escaping; in the case of alimentary CS, the dog came up to the experimenter and assumed a begging posture. The correct instrumental CR previously established to the test CS appeared mostly in the presence of facilitatory components related to the training.

The performance of the general behavior related to the previously used procedures, indicates that the intermittent CS retains its classical (Pavlovian) connections with the unconditioned stimulus, even when it is tested in a different situation. However, *it does not retain the instrumental-type connections with the CS and UCS when it is applied in a different situation.*

A question can be asked, therefore, about the role of the environmental situation in the process of conditioning. The previous observations made by Konorski and Miller (1933, p. 114; Konorski

1967, pp. 359–360) showed that it is possible to establish an instrumental reaction to the situation alone. Later, conditioning to the situation alone was used by other investigators (Wyrwicka et al. 1959, 1960, cit. by Konorski 1967, p. 405). A more recent study by Wyrwicka and Chase (1994) on the conditioning of the presleep state to the experimental situation in cats is a further case of this kind of conditioning. The process of conditioning, therefore, is not limited to the single stimulus, but can occur also to a complex of permanent environmental stimuli.

The above described data support a view that the process of conditioning involves both the intermittent conditioned stimulus and the situation in which this stimulus appears. However, due to the experimental procedure which consists of giving the reinforcement *only* to the intermittent CS and never to the permanent complex of situational stimuli, the reaction to the situation alone ceases to appear. This is a *partial inhibition* of the conditioned reflex to the situation. In other words, *a differentiation takes place between the intermittent CS+situation and the situation alone*.

This does not mean that the CR to the situation alone completely disappears. It may still reappear in various circumstances, such as a sudden disturbance of the usual experimental procedure, etc. Moreover, the previously described experiments showed that the conditioned stimulus retains its acquired property of producing the CR only in the situation in which the training took place. When the CS was applied in a different situation, the CR was absent. This was especially evident in the case of instrumental conditioned reflexes.

Let us now discuss various authors' interpretations of the role of the situation, already mentioned in previous chapters. Konorski and Miller (1933, 1936, Konorski 1939, 1967) divided all stimuli into two groups. The permanent stimuli of the situation were called by them the "determining" stimuli, whereas the intermittent stimuli were called the "releasing" stimuli. The determining stimuli decide *which* reaction has to appear, while the releasing stimuli decide *when* it will happen. When studying their papers, the reader may draw a conclusion that Konorski and Miller did not regard the situation as a separate factor in conditioning. Rather they tried to eliminate the problem of the role of situation in conditioning, and in their usual procedure with conditioned reflexes, they quickly removed the intertrial appearances of the CR to the situation (by not reinforcing them) to concentrate only on the intermittent CS (cf. Konorski 1967, p. 359).

However, the problem of the situation persisted and in many cases seemed to be the decisive factor in behavior. One such instance is the study of Denisov and Kupalov (1933, Kupalov 1961), described in chapter 4. Briefly, in experiments on dogs, two kinds of illumination of the experimental chamber were used, full illumination in some sessions, and dim illumination in other sessions. When the conditioned reflex was trained in full light and then tested in dim light, its value remained at a high level during several successive sessions and only then decreased. Vice versa, after training in dim light with reflex testing in full light, its value remained low for the next several sessions. Kupalov (1961) suggested that this was an effect of conditioning to the experimental situation. According to Kupalov, however, the situation is not able to evoke the conditionwed reaction but can change the *tonus* of some parts of the brain. Therefore, he called the reflex to situation a *shortened* or *truncated* conditioned reflex.

However, this explanation cannot be accepted when confronted with facts such as the appearance of instrumental conditioned reactions to the situation only (cf. Konorski and Miller 1933, Konorski 1967, p.360; Wyrwicka et al. 1959, Wyrwicka and Chase 1994), or the absence of this reaction to the CS applied in a situation completely different from the situation in which this reaction was trained (cf. experiments described in chapter 5). However, through the use of special experimental procedure, the conditioned reflex to the situation became differentially inhibited. It was, therefore, considered a "shortened" or "truncated" conditioned reflex (Kupalov 1961).

Nevertheless, these terms cannot change the finding that in instrumental conditioning, *the conditioned stimulus retains its power only in the situation in which it was trained* (Wyrwicka 1958). When the CS is applied in a different situation, it does not produce the previously established instrumental reaction. Instead, it produces another instrumental CR, that which had been trained in this new situation. This phenomenon leads to the conclusion that, in fact, the *situational complex of stimuli dominates* over the intermittent stimulus.

As described in chapter 3, it was possible to obtain two different conditioned reactions when the same intermittent CS was applied in two different situations. Most data on this phenomenon (called "switching") were obtained in Asratyan's laboratory, mainly as a result of training. According to Asratyan's interpretation, switching results from the action of a specific permanent stimulus, e.g., the person conducting the experiment or the time of the day. The permanent stimulus was called by Asratyan the tonic stimulus which prepares the organism for a response to the phasic stimulus (i.e., intermittent CS). For instance, it was possible to obtain an alimentary conditioned reaction to a CS in the morning session, and a defensive conditioned reaction to the same CS in the afternoon session. Asratyan (1965) hypothesized that the switching phenomena result from the ability of the brain to adapt to environmental changes.

Although Asratyan's interpretation seems to be reasonable, the fact is that it does not take into account that the phenomenon of switching may be a result of *conditioning to the permanent situational* factor (such as the experimenter or the time of the day). All other interpretations can only add to this basic fact.

Similarly, even such a convincing interpretation as Konorski's theory of two kinds of stimuli, those that determine the kind of reaction and those which release the reaction (Konorski 1967), does not clearly speak about conditioning to the situation. It should be said that the determining stimulus (e.g., a cuff on the dog's leg), is simply another conditioned stimulus. Moreover, this situational conditioned stimulus dominates over the usual intermittent stimulus which only retains the role of an anonymous factor evoking the reaction.

The role of the environment in behavior was demonstrated in clinical studies of Lhermitte (1983, 1986 I and II). As described in chapter 4, his frontal lobectomy patients showed an extreme dependence on the situation, leading to an exaggerated imitative behavior (Lhermitte 1986,1) as well as to a kind of facilitation in responding related to each situation in which the patients found themselves (Lhermitte 1986, 11). This environmental dependence syndrome" (a term introduced by Lhermitte 1986, 11) could result from the frontal lobectomy. Nevertheless, the change in patients' behavior revealed one of the most important mechanisms of behavior normally obscured by other events.

It would also be interesting to quote here a behavior theory by Beritashvili (1971). His concept concentrates not on the reflexes but on *memory*. According to him, there are three kinds of memory: image memory, emotional memory, and conditioned-reflex memory.

Image memory is the highest form of memory found in higher vertebrates. According to Beritashvili, even a single presentation of a food object results in the formation of the image of this object in the brain; this leads to "image-driven" behavior each time the animal enters the room where the original presentation of food took place. *Emotional memory* originates in the fear related to a noxious agent. *Conditioned-reflex memory* (of a food object) presumably depends on the structural development of synaptic mechanisms.

Beritashvili concludes his theoretical considerations concerning memory in vertebrates by hypothesizing that each of the three forms of memory (image, emotional, and conditioned-reflex) gradually develop phylogenetically from fish to monkeys. All three kinds of memory are also present in man. The memory of man, however, differs specifically in that it includes "verbal-logical memory."

Although Beritashvili's concept of behavior is so different from the theoretical interpretations of other authors, some of his views are close to the problem of the role of situation discussed in this book. Beritashvili stresses the role of image memory as a reinforcing agent and its connections with the environment where the presentation of reinforcement occurs. In that way, the importance of the situation gains further support even from a distant point of view.

Special attention should be given to conditioning of the brain electrical activity (EEG) which occurs inside the body, i.e., in the *internal situation*. This type of conditioning was described in chapters 7 and 8. Examples of such conditioning of the electrical activity of the brain focused on self-producing the sensorimotor rhythm (SMR) related to the inhibition of epileptic seizures. The active production of SMR by epileptic patients indicates that this kind of behavior can result from learning, i.e., from the acquisition of an instrumental conditioned reflex. This reflex is evoked by the appearance in the brain of a complex pattern of motor excitation produced by specific physiological conditions (which normally lead to an epileptic seizure). The occurrence *of this pattern of excitation is a conditioned stimulus which evokes a specific instrumental conditioned reaction.* The conditioned reaction in this case is a self-produced specific physiological state of the brain which is expressed by an appearance of low-frequency, high amplitude EEG waves (SMR). The appearance of SMR shows that a state of motor inhibition has taken place including the *inhibition of epileptic seizures.*

This inhibitory reaction can be understood as an *avoidance* reaction, or, using the conditioned-reflex terminology, a *defensive instrumental conditioned reaction.* A question arises as to what is the reinforcement for this reaction. The answer is quite simple: *the reinforcement is a decrease of the motor excitation which prevents an epileptic attack.*

Here the whole conditioned act—stimulus, reaction, reinforcement—occurs totally inside the brain. Of course, the external environmental elements such as the room, recording devices, time of day, etc., also exist but their role in the process appears minimal. The main environment for the occurrence of this process is the *internal situation in the brain* in which the conditioned stimulus, the avoidance reaction, and the reinforcement, occur. It may be said that, in this case, instrumental conditioning to the internal situation takes place.

Let us add a few words of explanation of the term *internal situation*. It can be understood as a complex of memories of previous experiences of the organism. These memories may serve as an internal background for the final reaction. However, which of these memories are used as the main factors for the reaction has to be decided by the subject.

10

Summary and Conclusions

Chapter 1 summarizes the famous work of I.P. Pavlov. Using a single intermittent stimulus in the same unchangeable situation, isolated from external influences, Pavlov created the concept of the conditioned reflex as the fundamental unit of the activity of the brain. A conditioned reflex is an acquired function based on an unconditioned (inborn) reflex such as salivation to an object (including food) placed in the mouth, or lifting the leg, to pinching the skin of that leg. When an originally neutral stimulus, for example a tone, repeatedly precedes the unconditioned stimulus (also called the reinforcement) salivation or a defensive movement begins to appear to the tone and becomes a conditioned reaction.

Conditioned reactions, later called classical or Pavlovian conditioned reflexes, can be established practically to any of the previously neutral stimuli, such as various tones, visual objects, tactile and even some taste stimuli. The abundant experimental data obtained in Pavlov's laboratory led him to construct a theory of conditioned reflexes which included such functions as differentiation between stimuli, various forms of inhibition as well as mutual influences of some reflexes on the other reflexes, etc.

Chapter 2 deals with the motor conditioned behavior, which was not included in Pavlov's theory. According to Pavlov, the performance of a movement produces sensory impulses analogous to those produced by visual or acoustic stimuli. When these impulses are followed by an unconditioned stimulus (e.g., food in the mouth) they become a conditioned stimulus (CS). Pavlov, however, did not discuss the origin of the movement itself.

The mechanisms of the motor conditioned reflexes were first experimentally analyzed by two young scientists, J. Konorski and

S. Miller. They distinguished four varieties of these reflexes, two of them based on a positive unconditioned stimulus (such as food in the mouth) and two others on a negative unconditioned stimulus (such as pinching the skin). Konorski and Miller concluded that the motor reflexes differ from the Pavlovian classical reflexes in that the reinforcement depends on the performance (or inhibition) of a specific movement, whereas there is no such condition in the Pavlovian reflexes. In their terminology, the motor conditioned reflexes were called conditioned reflexes type II (CRII), while Pavlovian reflexes were given the term classical conditioned reflexes type I (CRI) (Miller and Konorski 1928, Konorski and Miller 1936, Konorski 1939, 1948,1967). Later, conditioned reflexes type I were called classical conditioned reflexes, while conditioned reflexes type II were called instrumental conditioned reflexes (Hilgard and Marquis 1940, Kimble 1961).

The observations of Konorski and Miller include the fact that the trained CRII appears not only to the intermittent conditioned stimulus (CS), but also to the situation in which it is applied, in the absence of the CS.

Chapter 2 also summarizes some results of studies in instrumental conditioning by other investigators such as Skinner (1938), Ferster and Skinner (1957), and others.

Chapter 3 deals with the phenomenon called "switching" which consists of an appearance of a conditioned reaction different than that originally trained to a definite CS. The first observations were made by Miller and Konorski (1928) who found that the instrumental reaction was dependent on fastening a band on the animal's leg or a change in the reinforcement rather than a change of the CS. Later studies of Fonberg (1967) showed that switching is possible not only between various reflexes with the same reinforcement, but also between the CRII with alimentary or defensive reinforcement, as well as between the CRII reinforced by either food or desirable electrical stimulation in the lateral hypothalamus (cf. Olds and Milner 1954, Olds 1962).

Chapter 3 also describes the results of extensive studies of switching by Asratyan and his associates (Asratyan 1951, 1961, 1965). In their studies the appearance of a different CR depended on factors such as different time of the session, different reinforcement, or using a band on the leg. Switching was also possible between the excitatory and the inhibitory CSs. Each of these stimuli represented a different permanent factor of the situation which had been connected through training with a different CR to the same intermittent CS.

Studies on switching included more recent experiments on human subjects with the defensive type of reinforcement; in these studies an additional CS, called a "contextual stimulus" (or "context"), preceding the usual CS, was applied (Kimmel and Ray 1978, Kimmel and Lachnit 1990, and others). The effects of the contextual stimulus were also studied on pigeons by Rescorla (1984). The studies on "context" resemble earlier experiments of Pavlov on a form of conditioning called by him "conditioned inhibition" (Pavlov 1960, Ch. 5).

Chapter 4 describes various studies on rats in which the situation turned out to be a potent factor producing CRs to some situational elements, such as the neuroleptic effects in the front of the room where injections of chlorpromazine were previously given (but not in the front of other rooms). Similarly, a specific conditioned circling was observed in the environment where apomorphine injections were previously given. Chapter 4 also includes a description of the situational effects on the behavior of human patients after frontal lobectomy.

Chapter 5 describes the systematic research on the role of the situational background and the intermittent stimulus in conditioning. When two instrumental CRs, with two different motor effects, one to an acoustic stimulus and the other to a visual stimulus, were trained in the same situation, each CS often evoked both CRs in the same trial or a CR trained to other CS. However, when each CR was trained in a situation completely different from the other situation, the correct CR was always obtained (see Table 5.1). When the CS was tested in the same or similar situation, it produced the correct CR. But when the CS was tested in a situation completely different from the original one, it did not evoke the correct CR but instead the CR which had been trained in that different situation to a different CS.

This includes the case of heterogeneous CRs. When the CS, originally trained to produce a defensive CR, was tested in the alimentary situation (in which alimentary CRs were trained to different CSs) it evoked some general defensive behavior (such as escape) but not the previously trained instrumental CR. The same was observed when an alimentary CS was tested in a defensive situation: it evoked a general alimentary posture such as begging, but not the previously acquired CR. However, an introduction of elements from the original situation (such as an object resembling the feeder) led to the performance of the correct CR. These observations show that the intermittent stimulus may retain its ability to evoke the acquired instrumental CR only in the situation in which it was originally trained (see Table 5.2).

Chapter 6 analyzes three examples of complex conditioning: detour, feeding, and presleep behaviors. The experimental analysis showed that each is a case of an instrumental conditioned reflex, occurring in a definite environment and acquired during a conditioning training, without the use of any intermittent conditioned stimulus. Each of these cases is complex and consists of a few simple motor activities. The detour behavior requires a bowl with food visible across an obstacle (fence). The act of eating also requires some preliminary learning, related to the kind of food to be consumed. Presleep behavior consists of a number of motor actions such as selecting a convenient place to lie down, assuming a specific position, closing the eyes, etc. All these motor actions depend on the situation in which they are supposed to occur.

Chapter 7 describes the conditioning of functions of various internal organs such as heart rate, increase or decrease salivation, vasoconstriction or vasodilation, and others.

Chapter 8 focuses on conditioning of electrical activity of the brain (EEG) in both animal and human subjects. Clinical studies showed that a certain EEG (decribed as the "sensorimotor rhythm" or the "EEG feedback") reduced or even prevented the epileptic seizures. This EEG pattern can be conditioned, i.e., produced by the patient, after a conditioning training. The EEG feedback conditioned reflex is evoked by a complex of stimuli deriving from the internal situation of the brain and is fully accomplished by an acquired EEG activity occurring in the brain. It is, therefore, an example of conditioning to the internal brain environment. An analysis of the clinical data has shown the beneficial effect of this specific EEG therapy.

Chapter 9 discusses the experimental and observational data described in previous chapters from a theoretical point of view.

Conclusions

The use of intermittent stimuli as the basis of the theory of conditioned reflexes by Pavlov was a basic achievement in explaining the function of the brain related to behavior.

Pavlov's theory, however, has left some questions unanswered, one of which is the role of the environmental background or situation in which the conditioned reflex (CR) occurs.

A number of observations show that the intermittent CS can retain its acquired properties only when it is applied in the situation where it had been established by the training. This is especially true for the instrumental type of conditioning. The instrumental type of conditioning cannot be evoked in a new situation that is completely different from the original one. Only when some elements related to the original situation are present, may the instrumental reaction appear.

However, in the case of very strong defensive unconditioned stimulus (for example, a painful electric shock to the skin), the related CS, used in the test situation, can evoke a general defensive reaction of classical type.

Each change in the situation can alter the performance of the CR.

The CR can be established not only to an intermittent stimulus but also to the complex of permanent situational stimuli. An example of CR to the situation is the appearance of the intertrial CRs, observed often in instrumental CR studies. These CRs to the situation are usually eliminated by non-reinforcing them.

The phenomenon of "switching" can be explained as a consequence of establishing the CR to each situation, which predominates over the effect of the intermittent stimulus.

The "truncated" conditioned reflex is nothing else than a conditioned reflex to the situation, partly inhibited by non-reinforcing it.

The facts of conditioning observed in pharmacological studies can also be explained by conditioning to the situation.

A special instrumental conditioning occurs in the therapy of epilepsy. It has been possible to establish a specific EEG pattern related to the suppression or prevention of seizures in epileptic patients. This fact can be understood as a case of establishing a defensive instrumental conditioned reflex to the internal situation.

The examples described in this book are only a few selected cases showing the crucial role of the situation in behavior. More studies on this topic are needed.

References

Anand, B.K. and Brobeck, J.R. (1951). Hypothalamic control of food intake in rats and cats. *Yale J. Biol. & Med.* 24:123–140.

Anokhin, P.K. (1961). Electroencephalographic analysis of cortico-subcortical relations in positive and negative conditioned reactions. *Ann. N.Y. Acad. Sci.* 92: 899–983.

Asratyan, E.A. (1951). Principle of trans-switching in conditioned reflex activity. *J. Higher Nerv. Activity* 1: 47–54 (Russian).

Asratyan, E.A. (1961). The initiation and localization of cortical inhibition in the conditioned reflex arc. *Ann. N.Y. Acad. Sci.*, 92: 1141–1159.

Asratyan, E.A. (1965). *Compensatory Adaptations, Reflex Activity and the Brain* (S.A. Corson, Transl.). Oxford: Pergamon Press.

Babb, M. and Chase, M.H. (1974). Masseteric and digastric reflex activity during conditioned sensorimotor rhythm. *EEG clin. Neurophysiol.* 36: 357–365.

Babkin, B.P. (1949, 1971). *Pavlov, A Biography*. Chicago and London: The University of Chicago Press.

Beritashvili, J.S. (1941). The comparative study of individual behavior in dogs, rabbits and hens. *Transactions of the Beritashvili Physiological Institute*, 4:229.

Beritov (Beritashvili), J.S. (1948). Fickleness of individual reflexes under influence of external stimuli. In *General Physiology of Muscular and Nervous System*. Ch. 11. Acad. Nauk USSR, Moscow.

Beritashvili, J.S. (1971). *Vertebrate Memory Characteristics and Origin*. (John S. Barlow, Transl.). New York, Plenum Press.

Beritashvili, J.S. and Akhmeteli, M. (1941). On the behavior of pigeons in overcoming an obstacle. *Transactions of the J. Beritashvili Physiological Institute*, 4: 248.

Black, A.H. (1967). Operant conditioning of heart rate under curare. Techn. Rep. No.12, October 1967. Dept. of Psychology, McMaster University, Hamilton, Ontario, Canada.

Brown, B.B. (1970). Recognition of aspects of consciousness through association with EEG alpha activity represented by a light signal. *Psychophysiology*, 6: 442–452.

Bykov, K.M. (1957). *The Cerebral Cortex and the Internal Organs*. Transl. and edited by W.H. Gantt. New York: Chemical Publ.

Caruso, L., Zozula, R., Davis, E.G. et al. (1988). Classical conditioning of sleep onset in humans. *Sleep Res.* 17:82.

Chase, M.H. and McGinty, D.J. (1970a). Modulation of spontaneous and re-flex activity of the jaw musculature by orbital cortical stimulation in the freely moving cat. *Brain Res.* 19: 117–126.

Chase, M.H. and McGinty, D.J. (1970b). Somatomotor inhibition and excita-tion by forebrain stimulation during sleep and wakefulness: orbital cortex. *Brain Res.* 19:127–136.

Chase, M.H. and Babb, M. (1973). Masseteric reflex response to reticular stimu-lation reverses during active sleep compared with wakefulness or quiet sleep. *Brain Res.*59:421–426.

Chase, M.H. and Harper, R.M. (1971). Somatomotor and visceromotor corre-lates of operantly conditioned 12–14 c/s sensorimotor cortical activity. *EEG Clin. Neurophysiol.* 31:85–92.

Clemente, C.D. (1968). Forebrain mechanisms related to internal inhibition and sleep. *Conditioned Reflex*, 3:145–174.

Clemente, C.D., Chase, M.H., Knauss, T.K., Sauerland, E.K., and Sterman, M.B. (1966). Inhibition of a monosynaptic reflex by electrical stimulation of the basal forebrain or the orbital gyrus in the cats. *Experientia*, 22: 844–845.

Clemente, C.D. and Sterman, M.B. (1963). Cortical synchronization and sleep patterns in acute restrained and chronic cats induced by basal forebrain stimulation. EEG. *clin. Neurophysiol.* Suppl. 24172–187.

Clemente, C.D. and Sterman. M.B. (1967). Basal forebrain mechanisms for internal inhibition and sleep. In: *Sleep and Altered States and Conscious-ness*. Ch. Viii, pp. 127–147.

Clemente, C.D., Sterman, M.B. and Wyrwicka, W. (1963). Forebrain inhibi-tory mechanisms; conditioning of basal forebrain induced EEG synchroni-zation and sleep. *Exp. Neurol.* 7:404–417.

Corson, S.A. (1967). Cerebrovisceral theory: A physiological basis for psy-chosomatic medicine. *Internat. J. Psychol.* 4: 234–241.

Corson, S.A. and O'Leary-Corson, E. (1968). The effect of psychotropic drugs on conditioning of water and electrolyte excretion: Experimental researech and clinical implications. In: *Psychotropic Drugs in Internal Medicine Excerpta Medica International Congress Series*, 182: 147–164.

Cott, A., Pavloski, R.P. and Black, A.H. (1979). Reducing epileptic seizures through operant conditioning of central nervous system activity: proce-dural variables. *Science* 203: 73–75.

Denisov, P.K. and Kupalov, P.S. (1933). The magnitude of conditioned re-flexes in bright and obscure rooms. *Arch. Biol. Sci.* 35: 689. (Russian). Cit. *by* Giurgia 1989.

Descartes, Rene (1664). *L'Homme. Et un traite de la formation du foetus*. Ed. by Clerselier, Paris.

Di Cara, L.V. and Miller, N.E. (1968a). Long term retention of instrumentally learned heart-rate changes in the curarized rat. *Comm. Behav. Biol.* 2 (Part A): 19–23.

Di Cara, L.V. and Miller, N.E. (1968b). Instrumental learning of vasomotor response by rats: learning to respond differentially in the two ears. *Science* 159: 1485–1486.

Di Cara, L.V. and Miller, N.E. (1968c). Changes in heart rate instrumentally learned by curarized rats as avoidance responses. *J. Comp. Physiol. Psychol.* 65: 8–12.

Donhoffer, H. and Lissak, K. (1962). EEG changes associated with the elaboration of conditioned reflexes. *Ann. physiol. Acad. Sci. Hungarica* 21: 249–255.

Doty, R.W. and Giurgea, C. (1961). Conditioned reflexes established by coupling electrical excitation of two cortical areas. In: *Brain Mechanisms and Learning* (A. Fessard, R.W. Gerard and J. Konorski, eds.). Blackwell Scient. Publ., pp. 133–151.

Eikelboom, R. and Stewart, J. (1979). Conditioned temperature effects using morphine and the unconditioned stimulus. *Psychopharmacology* 61: 31–38.

Ellertsen, B. and Klove, H. (1976). Clinical application of biofeedback training in epilepsy. *Scand. J. behav. Therapy.* 5: 133–144.

Evans, D.R. and Bond, I.K. (1969). Reciprocal inhibition therapy and conditioning in the treatment of insomnia. *Behav. Res. Ther.* 7a: 323–325.

Ferster, CV.B. and Skinner, B.F. (1957). *Schedules of Reinforcement.* New York: Appleton-Century-Crofts.

Finley, W.W., Smith, H.A. and Etherton, M.D. (1975). Reduction of seizures and normalization of the EEG in a severe epileptic following sensorimotor feedback training. Preliminary Study. *Biol. Psychol.* 2: 189–203.

Fishel, W. (1933). Uber bewahrende und wirkende Gedachtnisleistung. *Biol. Zbl.* 53: 349.

Fonberg, E. (1961). On the transfer of two different defensive conditioned reflexes type II. *Bull. Pol. Acad. Sci. Cl.II,* 9 : 4 7 –4 9 .

Fonberg, E. (1967). The motivational role of the hypothalamus in animal behavior. *Acta Biol. Exper. (Warsaw)* 27: 303–318.

Fonberg, E. (1969). The role of the hypothalamus and amygdala in food intake, alimentary motivation and emotional reactions. *Acta Biol. Exper. (Warsaw)* 29: 335–358.

Galef, B.G.,Jr. (1978). Differences in affiliative behavior of weanling rats selecting eating and drinking sites. *J. Comp. Physiol. Psychol.* 92: 431–437.

Galef.B.G.,Jr. and Clark, M.M. (1972). Mother's milk and adult presence: two factors determining initial dietary selection by weanling rats. *J. Comp. Physiol. Psychol.* 78: 220–225.

Galef, B.G.,Jr. (1977). Social transmission of food preferences: an adaptation for weaning in rats. *J. Comp. Physiol. Psychol.* 91: 1136–1140.

Gantt, W.H. (1944). *Experimental Basis for Neurotic Behavior.* New York: Hoeber.

Gantt, W.H. (1960). Cardiovascular component of the conditioned reflex to pain, food and other stimuli. *Physiol. Rev.* 40, suppl. 4: 266–291.

Gantt, W.H. and Hoffmann, W.C. (1940). Conditioned cardio-respiratory changes accompanying conditioned food reflexes. *Am. J. Physiol.* (Proc.): 129: 360–361.

Gastaut, H. (1958). Some aspects of the neurophysiological basis of conditioned reflexes and behavior. In: *Neurological Basis of Behavior* (G.E. Wolstenholme and C.M. O'Connor, eds.), pp. 255–276. Boston: Little, Brown and Co.

Giurgea, C.E. (1970). Central excitability tonus and its modulatory influence upon a neuroleptic activity. In: Modern *Problems of Pharmacopsychiatry. The Neuroleptics.* Vol. 5: 91–94. Basel, Karger.

Giurgea, C.E. (1989). Kupalov's concept of shortened conditional; reflexes: psychophysiological and psychopharmacological implications. *Pavl. J. Biol. Sci* 24: 81–89.

Harper, R.M. (1973). Relationship of neuronal activity to EEG waves during sleep and wakefulness. In: M.I.Phillips (ed.) *Brain Unit Activity during Behavior*, pp. 130–154. Springfield, IL.: C.C. Thomas.

Harper, R.M. and Sterman, M.B. (1972). Subcortical unit activity during conditioned 12–14 Hz sensorimotor rhythm in cats. *Federation Proceedings* 31:404.

Higginson, G.D. (1926). Cit. by Thorpe 1956.

Hilgard, E.R. and Marquis, D.G. (1940). *Conditioning and Learning.* New York: Appleton-Century-Crofts.

Hsiao, H.H. (1929). Cit. by Thorpe 1956.

Izquierdo, I., Wyrwicka, W., Sierra, G. and Segundo, J.P. (1965). Etablissement d'un reflexe de trace pendant le sommeil natural chez le chat. *Actual. Neurophysiol.* 6: 277–296.

Jasper, H.H. and Ajmone-Marsan, C. (1960). *A Stereotaxic Atlas of the Diencephalon of the Cat.* Ottawa: The National Research Council of Canada.

John, E.R., Leiman, A.L. and Sachs, E. (1961). An exploration of the functional relationship between electroencephalographic potentials and differential inhibition. *Ann. N.Y. Acad. Sci.* 92: 1160–1182.

Kamiya, J. (1962). Conditioned discrimination of the EEG alpha rhythm in humans. Paper presented at the Meeting of the *Western Physiological Aassociation.* San Francisco, April 1962.

Kamiya, J. (1968). Conscious control of brain waves. *Psychology Today* 1:57–60.

Kaplan, B.J. (1975). Biofeedback in epileptics: equivocal relationship of reinforced EEG frequency to seizure reduction. *Epilepsia* 16: 477–485.

Khananashvili, M.M. (1983). *The Pathology of the Higher Nervous Activity* (Russian). Moscow: Edit. Medizina.

Kimble, G.A. (1961). *Hilgard and Marquis Conditioning and Learning.* New York: Appleton-Century-Crofts.

Kimmel, H.D. and Ray, R.L. (1978). Transswitching: Conditioning with tonic and phasic stimuli. *J. Exper. Psychol.: General* 107: 187–205.

Kogan, A.B. (1960). The manifestations of processes of higher nervous activity in the electrical potentials of the cortex during free behavior of animals. *EEG clin. Neurophysiol.* Suppl. 13: 51–64.

Köhler, W. (1948, orig.1925). *The Mentality of Apes.* London.

Konorski, J. (1939). Principles of cortical switching. Przeglad Fiziol. Ruchu 9: 191–242. (Polish; a French summary).

Konorski, J. (1948). *Conditioned Reflexes and Neuron Organization.* London: Cambridge University Press.

Konorski, J. (1967). Integrative Activity of the Brain. *An Interdisciplinary Approach.* Chicago: University of Chicago Press.

Konorski, J. and Lawicka, W. (1959). Physiological mechanisms of delayed reactions. *Acta Biol. Exp. (Warsaw)* 19: 175–197.

Konorski J. and Miller S. Conditioned reflexes of the motor analyzer. *Trudy Fisiol. lab. Pavlova.* 6: 119–298 (Russian).

Konorski, J. and Szwejkowska, G. (1950). Chronic extinction and restoration of conditioned reflexes. 1. Extinction against the excitatory background. *Acta Biol. Exp. (Warsaw)* 15: 135–170.

Konorski, J. and Wyrwicka, W. (1950). Transformation of conditioned reflexes of the first type into conditioned reflexes of the second type. *Acta Biol. Exp. (Warsaw)* 15: 193–204.

Kozak, W., Macfarlane, W.V. and Westerman, R. (1962). Long-lasting reversible changes in the reflex responses of chronic spinal cats to touch, heat and cold. *Nature (Lond.)* 193: 171–173.

Kozak, W. and Westerman, R. (1966). Basic patterns of plastic change in the mammalian nervous system. In: *Nervous & Hormonal Mechanisms of Integration, XXth Symposium of the Society for Experimental Biology.* 20: 509–544.

Krasnogorsky, N.I. Cit. by Pavlov 1951.

Kuhlman, W.N. (1978). EEG feedback training: enhancement of sensorimotor cortical activity. *EEG clin. Neurophysiol. 45:* 290–294.

Kuhlman, W.N. and Allison, T. (1977). EEG feedback training in the treatment of epilepsy: some questions and some answers. *Pavlovian J. Biol. Sci.* 12: 112–122.

Kuhlman, W.N. and Kaplan, B.J. (1979). Clinical applications of EEG feedback training. In R.J. Gatchel and K.P. Price (eds). *Clinical Applications of Biofeedback: Appraisal and Status.* New York: Pergamon.

Kupalov, P.S. (1961). Some normal and pathological properties of nervous processes in the brain. *Ann. N.Y. Acad. Sci.* 92: 1046–1053.

Lachnit, H. (1986). Transswitching and contextual conditioning. Relevant aspect of time. *Pavlovian J. Biol. Sci.* 21: 160–172.

Lachnit, H. and Kimmel, H.D. (1990). Contextual conditioning. A comparison of Eastern and Western views. *Pavlovian J. Biol. Sci.* 25: 174–179.

Lantz, D. and Sterman, M.B. (1992). Neuropsychological prediction and outcome measures in relation to EEG feedback training for the treatment of epilepsy. In: Thomas L. Bennett (Ed.) *The Neuropsychology of Epilepsy.* New York: Plenum Press.

Lawicka, W. (1959). Physiological mechanism of delayed reactions. II. Delayed reactions in dog and cats to directional stimuli. *Acta Biol. Exp. (Warsaw)* 19:199–219.

Lhermitte, F. (1986). Human autonomy and the frontal lobes. Part II. Patient behavior in com-lex and social situations. The "environmental dependency syndrome." *Ann. Neurol.* 11: 335–343.

Lorenz, K. (1932). Betrachtungen uber das Erkennen der arteigenen Triebhandlungen der Vogel. *J. Ornithol.* 80:50.

Lorenz, K. (1939). Vergleichende Verhaltungsforschung. *Zool. Anz.*12, Supplementband, p.60.

Lubar, J.F., Shabsin, H. Natelson, S.E., Holder, G.S., Whitsett, S.F., Pamplin,

W.E. and Krulikowski, D. (1981). EEG operant conditioning in intractable epilepsy. *Archives Neurol.* 38: 700–704.

McGinty, D.J. and Sterman, M.B. (1968). Sleep suppression after basal forebrain lesions in the cat. *Science* 160: 1253–1255.

Melchior, Chr.L. (1988). Environment-dependent tolerance to ethanol produced by intracerebroventricular injections in mice. *Psychopharmacology* 96: 258–261.

Miller, N.E. (1969). Learning of visceral and glandular responses. *Science* 163: 434–445.

Miller, N.E. and Carmona, A. (1967). Modification of visceral response, salivation in thirsty dogs, by instrumental training with water reward. *J. Comp. Physiol. Psychol.* 63: 1–6.

Miller, N.E. and Di Cara, L.V. (1967). Instrumental learning of heart-rate changes in curarized rats: shaping and specificity to discriminative stimulus. *J. Comp.Physiol. Psychol.* 63: 12–19.

Miller, N.E. and Di Cara, L.V. (1968). Homeostasis and reward: T-maze learning induced by manipulating antidiuretic hormone. *Am. J. Physiol.* 215: 684–686.

Miller, S. and Konorski, J. (1928). Sur une form particuliere des reflexes conditionnels. *C.R. Soc. Biol.* (*Paris*) 99: 1155–1157.

Murrin, M.R. and Kimmel, H.D. (1986). Determinants of tonic and phasic reactions in transswitching. *Pavlovian J. Behav. Sci.* 21: 117–123.

Newmann, T. (1978).

Nowliss, D.P. and Kamiya, J. (1970). The control of electroencephalographic alpha rhythms through auditory feedback and the associated mental activity. *Psychophysiology* 6: 476–484.

Olds, J. (1962). Hypothalamic substrates of reward. *Physiol. Rev.* 42: 554–604.

Olds, J. and Milner, P. (1954). Positive reinforcement produced by electrical stimulation of septal area and other regions of rat brain. *J. Comp. Physiol. Psychol.* 47: 419–427

Olds, J. and Olds, M.E. (1961). Interference and learning in paleocortical systems. In *Brain Mechanisms and Learning* (A. Fessard, R.W. Gerard, J. Konorski and J.F. Delafresnaye, eds.), pp. 153–187.

Pavlov, I.P. (1949). *Pavlovskie Sredy* 1949, p. 574. (Russian).

Pavlov, I.P. (1951). Fiziologicheskii mechanism tak nazyvanych proizvolnykh dvizenii. In: I.P.Pavlov's *Polnoe Sobranie Sochinenii*, Vol. 3, Book 2, 2nd edition, enlarged, pp. 315–319. Moscow and Leningrad: Academia Nauk SSSR.

Pavlov, I.P. (1960, first printed 1927). *Conditioned Reflexes*. New York: Dover Publications.

Poser, E.G., Fenton, G.W. and Scotton, L. (1965). The classical conditioning of sleep and wakefulness. *Behav. Res. Ther.* 3: 259–264.

Poulos, C.X. and Hinson, R. (1982). Pavlovian conditional tolerance to haloperidol catalepsy: Evidence of dynamic adaptation in the dopaminergic system. *Science* 218: 491–492.

Rescorla, R.A. (1984). Associations between Pavlovian CSs and context. *J. Exper. Psychol.: Animal Behavior Processes*. 10: 195–204.

Roginsky G.Z. and Tikh (1956). Roundabout ways in animals (Summary). *Problems of the modern physiology of the nervous and muscle systems.* Tbilisi, p. 384.

Rosenfeld, J.P., Rudell, A.P., and Fox, S.S. (1969). Operant control of neural events in humans. *Science* 165: 821–823.

Roth, S.R., Sterman, M.B. and Clemente, C.D. (1967). Comparison of EEG correlates of reinforcement, internal inhibition and sleep. *EEG clin. Neurophysiol.* 23: 509–520.

Sakhiulina, G.T. Cit. by Asratyan 1965.

Schiller, P. von (1942). Unwegversuche an Elritzen. *Z. Thierpsychol.* 5:101.

Seifert, A.R. and Lubar, J.F. (1975). Reduction of epileptic seizures through EEG feedback training. *Biol. Physiol* 3: 157–184.

Sherrington, C.S. (1929). Sherrington's Ferrier Lecture. *Proc. Roy. Soc. B*, Vol. CV, p.332. Cit. by Konorski 1948, p. 72.

Sherrington, C.S. (1947). *Integrative Action of the Nervous System.* Cit. by Konorski 1948, p.72.

Skinner, B.F. (1938). *The Behavior of Organisms: an Experimental Analysis.* New York: Appleton-Century.

Soltysik, S. (1975). Post-consummatory arousal of drive as a mechanism of incentive motivation. *Acta Neurobiol. Exp.* 35: 447–474.

Sterman, M.B. (1981). Power spectral analysis of EEG characteristics during sleep in epileptics. *Epilepsia* 22 : 9 5 –10 6 .

Sterman, M.B. (1986). Epilepsy and its treatment with EEG feedback therapy. *Ann. Behav. Med.* 8: 21–25.

Sterman, M.B. and Clemente, C.D. (1962). Forebrain inhibitory mechanisms: Sleep patterns induced by basal forebrain stimulation in the behaving cat. *Exp. Neurol.* 6: 103–117.

Sterman, M.B. and Friar, L. (1972). Suppression of seizures in an epileptic following sensorimotor EEG feedback training. *EEG clin. Neurophysiol.* 33: 89–95.

Sterman, M.B. and MacDonald, L.R. (1978). Effects of central cortical EEG feedback training on incidence of poorly controlled seizures. *Epilepsia* 19: 207–222.

Sterman, M.B., MacDonald, L.R. and Stone, R.K. (1974). Biofeedback training of the sensorimotor EEG rhythm: Effect on epilepsy. *Epilepsia* 15: 395–417.

Sterman, M.B. and Shouse, M.M. (1980). Quantitative analysis of training, sleep EEG, and clinical response to EEG operant conditioning in epileptics. *EEG clin. Neurophysiol.* 49: 558–576.

Sterman, M.B. and Wyrwicka, W. (1967). EEG correlates of sleep. Evidence for separate forebrain substrates. *Brain Res* 6: 143–163.

Sterman, M.B., Wyrwicka, W. and Roth,S.R. (1969). Electrophysiological correlates and neural substrates of alimentary behavior in the cat. *Ann. N.Y. Acad. Sci.* 157: 723–739.

Stroganov, V.V. (1948). The effect of a change of the situation upon the higher nervous activity of the dog. *Trudy Fisiol. Lab. Pavlova* 13: 128–149.

Szwejkowska. G. (1950). The chronic extinction and restoration of conditioned reflexes. 2. The extinction against an inhibitory background. *Acta Biol. Exp. (Warsaw)* 15: 171–184.

Teyrovsky, V. (1930). A study of ideational behaviour in the Garden Warbler. *Pub. Fac. Sci. Univ. Masaryk*, No. 122.

Thorndike, E.L. (1898). Animal Intelligence. An experimental study of the associative processes in animals. *Psychol. Monogr.* 2, No. 8.

Thorpe, W.H. (1956). *Learning and Instinct in Animals*. London: Methuen.

Throwhill, J.A. (1967). Instrumental conditioning of the heart rate in the curarized rat. *J. Comp. Physiol. Psychol.* 63: 7–11.

Tolman, E.C., and Honzik, C.H. (1930). Cit. by Thorpe 1956.

Ungerstedt, U. (1976). 6–Hydroxydopamine induced degeneration of the migrostriatal dopamine pathway: The turning sundrome. *Pharmacol. Therap.* (B). 2:37.

Vatsuro, E.G. (1948). The conditioned reflex situation and its effect on learning of conditioned reflexes. *Trudy Fisiol. Lab. Pavlova* 13: 21–127. (Russian).

Windholz, G. and Wyrwicka, W. (1996). Pavlov's position toward Konorski and Miller's distinction between Pavlovian and motor conditioning paradigm. *Integr. Physiol. Behav. Sci* 31: 338349.

Wyler, A.R., Lockard, J.S., Ward, A.A. and Finch, C.A. (1976). Conditioned EEG desynchronization and seizure occurrence in patients. *EEG clin. Neurophytsiol.* 41: 501–512.

Wyler, A.R., Robins, C.A. and Dodrill, C.B. (1979). EEG operant conditioning for control of epilepsy. *Epilepsia* 20: 279–286.

Wyrwicka, W. (1952). On the mechanism of the motor conditioned reaction. *Acta Biol. Exp. (Warsaw)* 16: 131–137.

Wyrwicka, W. (1956). On the effect of experimental situation upon the course of motor conditioned reflexes. *Acta Biol. Exper. (Warsaw)* 17: 189–203.

Wyrwicka, W. (1958). Studies on the effects of the conditioned stimulus applied against various experimental backgrounds. *Acta Biol. Exper. (Warsaw)* 18: 175–193.

Wyrwicka, W. (1959). Studies on detour behavior. *Behaviour* 14: 240–264.

Wyrwicka, W. (1964). Unpublished experiments.

Wyrwicka, W. (1975). The sensory nature of reward in instrumental behavior. *Pavlovian J. Biol.* Sci. 10: 23–51.

Wyrwicka, W. (1982). Some aspects of the origins of food preferences. *Brain Res. Inst. Bull.* 6: 3–4, 10–11.

Wyrwicka, W. (1988). *Brain and Feeding Behavior*. Springfield, IL.: C.C.Thomas.

Wyrwicka, W. (1993). The problem of switching in conditional behavior. *Integr. Physiol. Behav. Sci.* 28: 239–257.

Wyrwicka, W. and Chase, M.H. (1972). The act of eating as an instrumental reaction rewarded by electrical stimulation of the brain. *Physiol. Behav.* 9: 717–720.

Wyrwicka, W. and Chase, M.H. (1994). Conditioning of brain stimulation-induced presleep behavior. *Physiol. Behav.* 56: 883–889.

Wyrwicka, W., Dobrzecka, C. and Tarnecki, R. (1959). On the instrumental reaction evoked by electrical stimulation of the hypothalamus. *Science* 130: 336–337.

Wyrwicka, W., Dobrzecka C. and Tarnecki, R. (1960). The effect of electrical stimulation of the hypothalamus on the conditioned reflexes, type II. *Acta Biol. Exper.* (*Warsaw*) 20: 121–136.

Wyrwicka, W. and Garcia, R. (1979). The effect of electrical stimulation of the dorsal nucleus of n. vagus on gastric acid secretion in cats. *Exp. Neurol.* 65: 315–325.

Wyrwicka, W., Sterman, M.B. and Clemente, C.D. (1962). Conditioning of induced EEG sleep pattern. *Science* 137: 616–618.

Wyrwicka, W. and Sterman, M.B. (1968). Instrumental conditioning of sensorimotor cortex EEG spindles in the waking cat. *Physiol. Behav.* 3: 703–707.

Zbrozyna, A. (1953). A phenomenon of non-identification of a stimulus operating against different physiological background in dogs. *Soc. Sci. Lodz, Sec. III*, No. 26 (Polish with an English summary).

Zuckermann, E. and Buffy, A. (1960). Protein-bounded acetylcholine output in cortical efferent blood after conditioned reflexes. *Exp. Neurol.* 2: 423–428.

Name Index

Subject Index